ECOLOGIES OF INVENTION

ANDY DONG
JOHN CONOMOS
BRAD BUCKLEY

First published in 2013 by Sydney University Press

© Individual contributors 2013

© Sydney University Press 2013

Sydney University Press
Fisher Library F03
University of Sydney NSW 2006 AUSTRALIA
Email: sup.info@sydney.edu.au

**National Library of Australia
Cataloguing-in-Publication entry**

Title: Ecologies of Invention — edited by
Andy Dong, John Conomos, Brad Buckley.

ISBN: 9781743323571 (paperback)
9781743322505 (ebook : epub)
9781743322512 (ebook : mobi)

Notes: Includes bibliographical references and index.

Subjects: Inventions — Discoveries in science — Essays.

Other Authors/Contributors: Dong, Andy, editor —
Conomos, John, editor — Buckley, Brad, editor.

Dewey Number: A824.008035

Cover and Slipcase Image:
Courtesy of Marc Newson Limited
Design: Lea Barnett, CampbellBarnett

Heading type set: Knarf Art Font 3
Text type set: TSTAR
Text and Cover stock: Pacesetter LaserPro
Slipcase stock: White Knight with black foil

ECOLOGIES OF INVENTION

ANDY DONG
JOHN CONOMOS
BRAD BUCKLEY

FOR
DOUGLAS ENGELBART
IN MEMORIAM
1925–2013

FOR
SEAMUS HEANEY
IN MEMORIAM
1939–2013

CONTENTS

ACKNOWLEDGEMENTS

The editors sincerely wish to thank the following people without whom this book project would not have been realised.

First and foremost, we thank Professor Jill Trewhella, the Deputy Vice-Chancellor (Research), and Professor John Redmond, Dean of the Faculty of Architecture, Design and Planning, at the University of Sydney for their support for ongoing collaboration between the researchers involved in the Sydney Research Networks Scheme (SyReNS) proposal—the Sydney Invention Studio. Their foresight in sponsoring a research compact to support this book project generously affirms the inventive achievements of researchers across the University and, more importantly, realises the University's commitment to horizontality across fields.

Susan Murray-Smith and Agata Mrva-Montoya of Sydney University Press contributed unwavering profession-alism and dedication in seeing our manuscript through from its conception to its realisation. Thanks to John Mahony for his consummate copyediting.

Two Sydney College of the Arts graduates contributed substantially to the book design. Lea Barnett (who graduated with a BVA in 1991) of CampbellBarnett produced a thoughtful, strategic, and inventive book design. Her book design concept pulled together one of the principal themes of the book, dialogic conversations between fields, and created a visual language to respond to the questions about inventiveness raised in the book. We gratefully thank Marc Newson (who graduated with a BA [Visual Arts] in 1984) for allowing us to use his exquisite image on the cover.

We are also especially grateful to all of our contributors who gave their time and expertise in helping us to materialise our editorial intentions and objectives. We span many fields and disciplines in the Faculty of Architecture, Design and Planning; Sydney College of the Arts; Sydney Conservatorium of Music; the University of Sydney Business School; School of Letters, Arts, and Media; and the Centre for Social Robotics—Australian Centre for Field Robotics, Faculty of Engineering and Information Technologies, and finding our common ground has been at once intellectually challenging and exhilarating. Many heartfelt thanks to all of you in sharing our concern for inventiveness and its impact on research in this unfolding century.

The editors would like to express our deep appreciation to our Project Officer, Dr Alexandra Crosby, for her professionalism, assistance, and multifaceted skills in thinking through all the details and strands of this project and bringing them to a unified, coherent whole.

Finally, Andy Dong wishes to convey his sincere thanks to his co-editors, Professor Brad Buckley and Associate Professor John Conomos, for their coaching and unerring critical voice. The opportunity to jointly interrogate, conjecture, and reflect on issues raised in this book is gratefully acknowledged. The ideas that flowed will continue to evolve as we continue to understand the phenomenon of inventiveness.

ANDY DONG
JOHN CONOMOS
BRAD BUCKLEY

FOREWORD

MARC NEWSON

My relationship with Sydney College of the Arts predates
the College's amalgamation with the University of
Sydney in 1990. During the 1980s I was a student in
the Sculpture and Jewellery Departments and started
making sculptures, objects and eventually chairs. I was
particularly interested in the Jewellery Department, not
because I wished to become a jeweller but because they
taught you how to make things, the importance of scale
and the value and weight of different materials. Perhaps
it was the freedom that art school offered rather than a
design school training that has allowed me to develop
a global approach to design and invention.

This book, *Ecologies of Invention,* considers
inventiveness rather than that much overused term
creativity. In this book inventiveness is explored across
a range of disciplines particularly but not exclusively
those outside engineering and the hard sciences,
which have traditionally been seen as the natural home
of invention in the university. Of course in a rapidly
changing world, where knowledge ebbs and flows
across what were once considered discreet boundaries,
this book makes an important contribution to our
understanding of that 'light bulb' moment of realisation
or invention that takes place across all disciplines,
allowing architects, artists, designers, musicians and
scientists to apply this inventiveness to innovation.

Importantly, *Ecologies of Invention* is structured
around the concept of scale, with chapters exploring
inventiveness from the micro- (personal capacities),
the meso- (social) and the macro- (spatial and network)
scales. This view of inventiveness across scales is what
differentiates this book from other ideas of innovation
or creativity and, as I have said on several occasions,
scale is central to my work as a designer. Although
the contexts of invention described in this book vary
significantly, the invention of a new watch, prototype
rocket or a new transit system, to name rather different
inventions, share similarities in terms of process,
cultural and social production.

Ecologies of Invention will make an important
contribution to our understanding of the role of
invention in our shared contemporary world.

INTRODUCTION

ANDY DONG

JOHN CONOMOS

BRAD BUCKLEY

INTRODUCTION

The fundamental objective of this book is to explore the role that inventiveness plays across all fields. It is customarily believed that invention is only applicable to a few fields such as medicine, engineering and the physical sciences. This belief unfortunately is evident across the higher education sector, government agencies, and other institutions, reflecting a general perception that invention originates from scientific discovery alone. Critically then this book describes the articulation of inventive capacities across disciplines with sensitivity to the personal capacities and social, spatial and network configurations that drive people to produce inventions.

What exactly then is inventiveness? This question goes back all the way to Plato's problem: How can we appear to have a body of knowledge more expansive than what we are exposed to through perception and action in the physical world? This is the central issue of invention. The traditional path to invention is purportedly independent investigation and discovery. Such assumptions demand serious scrutiny for many reasons, not the least of which is that they often become embodied into law and social policy. The contributions in this volume will show the shallow foundations of this and other key assumptions underlying invention. More so, the contributions address how to foster inventiveness within our own academic community of diverse scholars, artists, architects, designers, historians, engineers, doctors, physicists, chemists, lawyers and economists. The contributors to this volume grapple with this critical issue in our globalised world of media communications, geo-political, demographic, technological and pedagogic complexities and shifts on a basic level, amongst other related salient factors, to ponder the intricate reciprocal interaction of artistic practices and technological and scientific developments.

Simply put, is an artist, architect, or designer an inventor? For us to scrutinise how inventiveness cuts across the entire domain of disciplines within a university—as this book clearly attempts to demonstrate—is to examine the elaborate connections between art, design, architecture, science and engineering and not to argue that it is only located within the traditional cluster of the physical and natural sciences.

Specifically then, the book articulates inventive capacities across disciplines with a sensitivity to the social, spatial and network configurations within which each of the inventors who contribute to this book are embodied, and describes ecological connections across these scales, which would otherwise be invisible. Therefore, inventiveness is a multi-scale issue, containing 'components' from the micro- (personal capacities), the meso- (social), and the macro- (spatial and network) scales. This view of inventiveness across scales is what differentiates our book. Although the contexts of invention described in this book vary significantly, we believe that the invention of a new style of poetry and the invention of a new consumer service, to name two inventions at opposite ends of the spectrum, share regularities in terms of process and cultural production.

Some of the key questions that this book addresses include: What critical patterns and regularities recur in inventors' capacities? What architectural, social and spatial configurations foster invention? What behaviours and forms do online and physical networks of invention actually take? What important factors reliably predict rates of inventiveness?

These research questions are of broad importance across disciplines beyond those represented in the book. Economists model inventiveness as an 'absorptive capacity', a firm's ability to assimilate and exploit external knowledge as a precursor to innovation in new ideas.[1] In neuroscience, inventiveness is seen as a byproduct of the network architecture of the human brain, which allows the activation of mental representations that are not part of ambient reality.[2] Finally, anthropologists study the evolution of humans through a comparative analysis of artifacts created by early humans and our closest species relatives, the great apes, to theorise about the evolution of cognitive abilities for inventiveness.[3] Ultimately, if we are to understand intelligence in humans and what constitutes personhood, we need to consider how it is that humans produce culture, which Holloway[4] defined as the imposition of arbitrary form on the environment. This imposition of arbitrary form can only happen through the invention of instruments to inscribe our identities.

Hitherto, inventiveness has received limited attention as an object of study. Whereas researchers across the academy could easily describe their research methods, and research methods are intensively documented and subjected to scrutiny, the act of invention is largely treated as indescribable or taken for granted as miraculous moments of serendipity that elude codification. This is academically problematic given that research discoveries and inventions are largely the same; research discoveries bring advances of a conceptual nature whereas inventions bring advances of an applied nature.[5]

Unfortunately, economic rationalist and intellectual property motives have burdened discourse around invention, leading many researchers whose intellectual tradition is based on inventiveness to retreat to the concepts of creativity and innovation. Basically neither of these concepts is particularly helpful when trying to describe the relatively known intellectual traditions of invention. Neither of these concepts adequately characterises the intellectual tradition of invention. Inventiveness subsumes creativity and demands cumulative knowledge-building across intellectual fields. It is also fundamentally the precursor to innovation.

Ecologies of Invention is the first edited collection of essays that brings together writers, several of whom are scholars of international standing, and a number of current academics and graduates of the University of Sydney to focus on and examine how invention is impacting on and changing our culture and society. With our selection of contributors we have decided that there should be a clear balance between the conceptual and methodological, and the empirical chapters. We have chosen a number of scholars who are widely acknowledged for their national and international standing in their respective fields. The book's concept of invention extends beyond the essays to the cover image and its design demonstrating the depth of inventiveness across the University.

Consequently, in order to do justice to such a significant topic, it is imperative that we discuss a wide array of different creative, intellectual and scientific approaches, and, in so doing, appreciate how some of these approaches in themselves may be contradictory but essentially are equally necessary, possible and legitimate. Invention in itself is a very multifaceted phenomenon that encapsulates, at any given historical moment of critical reflection, many different diverse theoretical approaches, intellectual traditions, and methodologies. However, what animates this book's underlying conceptual architecture is the salient belief that invention itself embodies a number of disparate factors that must come together in order for it to exist. Whatever the discipline, for invention to take place, it must be located in the mutating complex interrelationship of culture, technology and science. Within this important axis of human creativity, curiosity, scientific, and technological pursuit, the arts do play a role of considerable importance. Invention, contrary to public perception, does figure in the humanities and the creative arts as much as it does elsewhere in the sciences. This is becoming clearer by the day.

The late US mathematician Norbert Wiener, one of the first to contribute to this particular view of invention, emphasised how certain factors must surface first and collide to produce invention. Wiener's Promethean intellect and existential wisdom always underscored the role that individuality plays in invention: 'Thus one of the purposes of the present book is to make a proper assessment of the individual element in invention and discovery and of the cultural element.'[6] This perspective of invention is so often forgotten in a university's foundational pedagogic ecology and research praxis, and needs to be cogently foregrounded in any present and future discussion of such a topic. Too much is at stake otherwise.

The importance of inventiveness to Australia, which along with many other Western nations is facing a declining or exhausted manufacturing base, has become urgent. This has created a situation where countries are scrambling to maintain an edge in the global economy through the production of new knowledge and invention. In the US, inventiveness—though the term 'innovation' is generally used when discussing these issues—is widely considered to be the product of science, technology, engineering and maths and is known by the acronym STEM.

However, Rhode Island School of Design (RISD) president John Maeda and other academics at RISD have led the way with a groundbreaking initiative that places art and design at the centre of the equation transforming STEM into STEM + Art = STEAM. According to Maeda, STEAM 'addresses creative problem solving, the translation of complex data for broad audiences through visualisation, and how to bring ideas to market through design'.[7]

In many ways this book attempts to explore Maeda's concept of STEAM by contesting the prevailing hegemonic and distorted belief that invention can only take place in the sciences and not in any other discipline as such. This is lamentably unhelpful if we wish for universities to adequately realise their innovation potential in terms of creative industries, economy, productivity, and research and development. Further, and critically, not only would such a more democratic, empirically realistic and pragmatic view of invention advance knowledge, government innovation policies and technology, it would also considerably enhance cultural, economic and social benefits and outcomes of our national creative sector.

In a critical sense, this book wishes to bring to light the 'hidden' potential of invention and innovation that resides in those disciplines outside the customary ones of science, medicine, engineering and technology. In other words, it is quite timely now to suggest that the creative arts, humanities and social sciences are (when it comes to invention) more than just the tokenistic contributors they are often regarded as. In fact, when one examines the subject more critically from an informed position sceptically— and adopting a global reflexive approach to its theoretical, cultural, empirical and methodological complexities and realities—then one realises how prominent these disciplines really are.

It is more than 50 years ago that the novelist and scientist C. P. Snow coined the influential term 'two cultures' to describe the different world views of understanding reality by scientists and the artistically creative.[8] Snow famously spoke of a hostile dislike between the literary intellectuals and scientists of his time. In fact, since Snow's time, the idea of innovation has become categorically aligned with modern science. This problematic view is still unfortunately evident as the totemic norm in a university's disciplinary ecology and governance. Therefore, it is crucial at this historical juncture to deploy a more inclusive multifaceted framework and appreciation of the subject, illustrating the value and role that the creative arts, humanities and social sciences play. As media and communications theorist Stuart Cunningham has recently persuasively argued, following Ian Miles and Lawrence Green's 2008 study of Britain's innovative enterprises of advertising, independent broadcasting, games and product design, the innovation of the creative industries including the 'new' humanities disciplines like media, cultural and communication studies, including the creative arts, is 'hidden' in contemporary academic, cultural and social life.[9]

There needs to be more elaborate thinking about invention, universities, and government innovation policy. As Cunningham rightly argues, and as this book rigorously seeks to address, we need to think that invention and innovation are 'far more than white-lab-coat science, and high-value service industries, as are found in the creative sector, are where a large proportion of incremental and process innovation happens'.[10] What is required now more than ever is the recognition that, since Snow's 1959 *Two Cultures* debate, what we are encountering today is that the traditional dualistic split between the arts and the sciences has been transformed into an elaborate and subtle open-ended spectrum of aesthetic, cultural, theoretical and technological ideas that cut across disciplines when it comes to invention. This is something that the literary historian and critic Stefan Collini minimally refers to in his 1998 introduction to Snow's benchmark publication when he acknowledges the 'Rubik's-cube' recombinant shadings of this spectrum:

> Reflection on this point should do more than simply soften Snow's original polarity into a more continuous spectrum … We need, rather, something like multidimensional graph paper in which all the complex parameters which describe the interconnections and contrasts can be plotted simultaneously.[11]

Cunningham argues therefore that, when discussing innovation in the context of industry, policy and the creative sector, what is urgently needed is to recalibrate our thinking concerning the creative industries in a non-totalising, non-dualistic fashion by being aware of how much of contemporary life in all of its spheres is shaped by the new proliferation of digital and virtual technologies. Furthermore, we need to ask ourselves (theoretically and empirically) who benefits from a university's ecology of disciplines that adheres to a more traditional definition of invention located only in the sciences?

As media theorists Dieter Daniels and Barbara U. Schmidt rightly point out in their probing analysis of the artist as inventor and vice versa, Wiener's groundbreaking theory of cybernetics introduces to us a much more critically poetic, sceptical and sophisticated understanding of the role that the creative intellect plays in combination with a whole cluster of different variables such as the cultural context, techniques, materials, social class, economics and politics.[12] Wiener, who profoundly understood the dialectical complexities and tension between the arts and science, invented the concept of cybernetics, which, in its subsequent popularised manifestation as the 'third culture', critically corresponded, in Daniels' and Schmidt's words, 'today to the widespread propagation of digital phenomena in the culture of everyday life'.[13] This propagation, as these two authors opine, gave rise to a pervasive proliferation of aesthetic and technical hybridisation shaping our creative work processes and cultural practices across the whole spectrum of disciplines when it comes to a fundamental understanding of invention in and outside the academy.

Ecologies of Invention consists of seven chapters that discuss, in their respective transdisciplinary theoretical and methodological voices, various concepts, contexts and research strategies dealing with the complexities of invention in terms of knowledge production, experimentation and research-praxis as encountered mostly across the disciplinary spectrum at the University of Sydney. At all times, there is a critical consensual effort amongst the various contributors to try to carefully analyse the many shifting intricacies of invention as they are encountered by the various contributors in their specific disciplines of research, teaching and technologies. Invention is, at all times, carefully differentiated from creativity and innovation.

Each book chapter is divided into two parts which essentially are dialogic and interrelated to each other concerning invention in terms of its different critical capacities, sensibilities and social, cultural and spatial configurations in a particular discipline. This approach has been used to describe the aesthetic, cultural, experimental, theoretical, spatial, and technological complexities of invention as it applies to that particular discipline in question. Thus, the first part is an introductory theoretical essay that contextualises the following part, which in effect is a case study of invention at work in the University and beyond.

One of the most important editorial strategies of this approach is to present to the reader an enormous breadth, complexity, multidimensionality, and transdisciplinary nature of invention that characterises the University's creative, intellectual, pedagogic and technological ecology. This is crucially unprecedented as the book globally showcases invention across the University. In so doing, *Ecologies of Invention* strives to bridge together all the so far disparate contexts, elements, disciplines and technologies of invention as a continuing dialogue of possibilities and new 'grammars of creation'.[14]

In chapter one, Andy Dong's multifaceted contextualising essay 'Discourses of Intervention: A Language of Invention' deftly and succinctly maps out some of the more current vital and compelling views relating to the emerging unexplored surface contact between the disciplines in certain situations of invention. Dong argues by analogy borrowed from psychotherapy—the concept of 'intervention' in particular— that in architecture and engineering it signifies the act of reinventing space and form by rejecting bonds to the history of space and place. Thus in invention what we have is, in fact, an 'intervention' that in various fields of knowledge unmaps and destabilises official knowledge so that notions of canonical definitions, specialised methods and established identities representing a field of knowledge are eschewed in favour of uncircum-scribed potentials.

Dong's guiding expertise in design studies, specifically on what constitutes design knowledge and relatedly the causal significance of the processes and structures of design knowledge production on design-led innovation, primarily has shaped (in conjunction with Brad Buckley, an artist, urbanist and polemicist, renowned for his 'post-medium' expertise in installation, theatre and performance) the book's characteristic conceptual, formal and thematic concerns.

Grounded in social realism and Basil Bernstein's theories of vertical and horizontal discourses, Dong's illuminating essay addresses the structural properties and principles of language in ways of speaking that open potentials for the creation of new realities.

Dong's essay focuses on how 'art-science' or 'art-technology' collaborations between artists and scientists need to, in order to produce new hybridised spaces of knowledge transmission and meaning-production that are central to the exigencies of the creative practice, break out of their own disciplinary boundaries and fixations. In short, as collaborators between disparate disciplines, for invention to take place they need, according to Bernstein's code theory, to conceptually, dialogically and performatively engage in constructing the new object they are creating. When there is a clash of disciplinary codes happening, then collaborators in their discourses of invention are essentially able to weaken disciplinary boundaries and control of permissible knowledge and introduce the necessary horizontal structures of knowledge that are mandatory for a new specialised language of new questions, new claims and voices to surface.

This is followed by Petra Gemeinboeck and Rob Saunder's contribution 'Inventing Cultural Machines', which adroitly investigates the inventive processes of their art and technology collaboration that are particularly situated between the fields of artificial creativity and experimental art. And in so doing they perfectly illustrate the main critical points of Dong's interdisciplinary design-oriented thesis using Bernstein's code theory of 'discourses of invention' to account for the creative, intellectual, performative and technological complexities of invention through collaboration between the arts and sciences.

For Gemeinboeck and Saunders, artistic inventiveness does not principally only concern the production of an original 'artefact', nor is it the outcome of the experimental technical development in their collaborative work. But rather they see it as a strategic intervention into cultural and social discourse, whereas the artwork is its aesthetic materialisation and the technology is at once the medium for intervention and a complicated actor.

In relation to their robotic artwork *Zwischenräume* (2010–12), the authors have attempted to tease out the composite nature of inventiveness in the dynamic interplay of several contexts and positions, including the cultural, social and technological context of their art-making and the artwork itself. Also, invention for them indubitably is not only a matter of the dialectical concerns, methods and tensions of interdisciplinary collaboration but inevitably also of a collision of world views that Dong himself spoke of in his preceding contribution. It is also equally a matter of constructing new knowledge production and utilising multiple (human and non-human) actors salient to their art–technology collaborations. Finally, invention for Gemeinboeck and Saunders involves other indispensable factors such as machine autonomy and the role of the inventor and public exhibitions as testing grounds.

Dan Lovallo's succinct contribution 'The "Character" and the "Algorithm": An Essay on Technology and Art' is primarily an overview of four important new works by several artists and researchers working at the University of Sydney. Its main critical focus is, however, on two new installations recently exhibited at Artspace Visual Art Centre in Sydney in May of 2013.

The first installation is Mari Velonaki's *The Woman and the Snowman,* which is comprised of two video projections. The first one is of a female android robot, Repliee Q2, dressed in a fine red gown located next to a tall wispy tree in the snow, by the renowned Japanese roboticist Professor Hiroshi Ishiguro from Osaka University. The second is of a snowman located in his natural environment, while there is a third non-anthropomorphic robot that more accurately represents a kinetic sculpture that moves in unison with a haunting score devised by composer Nikolas Doukas. This kinetic sculpture is also rotating and changing speed on several different levels to the score itself. It needs to be said that the sculpture also contains a few screens, which display scenes from a 1970s Greek television show and also a winter scene.

In Lovallo's estimation, the installation is in his words 'a sublime contemporary masterpiece', indicating Velonaki's diverse interests in Japanese cinema, performance art, robotics, kinetic sculpture, poetry, etc., and how the installation's defining inventiveness has been shaped by Velonaki's persuasive ability to engage high-calibre skilled colleagues.

The artists Petra Gemeinboeck and Rob Saunders, collectively known as 'robococo', collaborated on the work *Accomplice*. This exhibit consisted of four robots located on the gallery's walls communicating to each other both audibly and visually. But obviously these robots would also communicate with the participants in the gallery but not necessarily understanding what they were saying to each other. Each robot would communicate their presence by banging on their respective wall with an implement producing a hole in the wall. As the walls came apart, it was a case of seeing how these robots would behave in their unpredictable ways. This is essentially a performance installation using non-humanoid-shaped robots and, as an emergent system, the artists were concerned with experimenting with these robots so they could experiment with and imagine the future. In a certain sense, *Accomplice* evoked Jean Tinguely's famous self-destructing 1960 work *Homage to New York*.

Lovallo believes that one of the more interesting inventive attributes of robococo's work is their multifaceted capacity to utilise algorithms in such a new fascinating way by creating human characteristics such as curiosity in their robots. Fundamentally Saunders is breaking new ground by modelling curiosity algorithmically.

The author himself has in two recent algorithmic works, *Eucalyptus salmonophilia murray darling* and *Constellation 3—Exploded Julius 2013*, experimented innovatively with algorithms in order to produce witty, imaginative and compelling works. With the former, Lovallo used a branching algorithm to combine the branching of a tree and river basin. Significantly, the author used a 3D printer to create this work. As 3D printing costs fall, many more artists and technologists will be using this new revolutionary process of design and manufacturing. In fact, as the author reminds his readers, Boeing is already embarking on 'printing' their planes. And with the latter work Lovallo used an algorithm for the purpose of structuring simple materials like bottles, wires and coloured water into complex processes that would not have been achievable without the algorithm.

David Rye and Mari Velonaki's stimulating account of their cross-disciplinary collaboration as roboticists 'Art and Robotics—A Brief Account of Eleven Years of Cross-Disciplinary Invention', concisely delineates the aesthetic, disciplinary and technological complexities of creating works that speak of human–robot interaction deploying a basic interest in novel human–machine interfaces as well. Velonaki's specific starting point of interest was in the production of haptic interfaces that explored different new models of interaction between the participant and an interactive kinetic object such as a robot. And for Velonaki's collaborators like Rye and Steve Scheding and Stefan Williams, all three roboticists, the direct challenge was to create a robot located in a public gallery space. Thus for Velonaki and her collaborators the inventiveness of their collaboration was to make a robot that was technologically intuitive in terms of its human–machine interface with the added challenge of ensuing that the robotic interactive artwork itself was also robust to unforeseen events in the gallery space.

The earlier stages of their collaboration were predicated on a cluster of issues concerning building bridges between different disciplines stemming from both shared and individual goals, trust as equal partners in the creative process, and the common realisation that, whatever one's disciplinary concerns and methodologies may be, there needs to be a fundamental acknowledgment that because of different educational and biographical circumstances wide cultural gaps are perhaps inevitable in such collaborative projects.

The authors' contribution engagingly describes the various challenging aesthetic and behavioural decisions that were made through their influential cross-disciplinary projects. *Fish-Bird, Fragile Balances, Diamandini* and *The Woman and the Snowman* are highly indicative of how the authors self-reflexively proceeded in their robotic and responsive artworks in relation to the complex human–machine interfaces deployed and the related bidirectional communication between gallery participants and the artwork in question and, in the case of female humanoid robot *Diamandini,* the haptic one-to-one human/robot interaction.

Their contribution clearly suggest the central importance in any cross-disciplinary collaboration of capitalising on the relevant individual and shared strengths of the participating artists, composers, roboticists and cinematographers, as in the case of their most recent multimedia/robotic work *The Woman and the Snowman.* Invention, for them, critically resides in clearly defining the effectiveness and experience of diverse interactions between gallery interactors and technological others. Invention in this situation, the authors argue, depended on their challenging mutual project creating new environments and experimental interfaces that would elicit new behaviours and new aesthetic experiences.

Kit Messham-Muir's chapter 'Melting into the Texture of Everyday Life' examines John Tonkin's acclaimed video installations of the last decade or so, in terms of how in our present digital and networked culture these works embody a new approach to interactivity that centres around ideas in keeping with contemporary social media and technological connectivity. Regarding such inventive works by Tonkin as *Selective Attention* and *Nervous System*, both of 2011, Messham-Muir clearly argues the case that this artist's works effectively question the more traditional 'keyboard and mouse' passive view of interactivity by inviting his audiences to be more participatory and collaborative in their encounters with his works. Interactive media art, for Tonkin, becomes emphatically more a process that is not interested in informational retrieval but one more of a creative and improvisational process.

In other words, the author posits the view that Tonkin's recent oeuvre, influenced to a considerable degree by a 2001 kinetic multimedia work *Welcome Space* at the new National Museum of Australia concerning Aboriginal and Torres Strait Islander cultures, is significantly socialised and 'mutualised' in its thinking about connectivity which is critically empathetic, culturally and socially, in its orientation. Given the huge impact that Web 2.0 technologies are having on our everyday life in changing knowledge ecologies from the old Gutenberg transmission mode of mediated knowledge to one of knowledge as a social participatory networking process of creative commons sharing, Tonkin's work is imaginatively and technologically inventive as it significantly facilitates a phenomenological person-to-person connectivism.

As John Tonkin clearly explains in his absorbing contribution 'On Building a Perceptual Apparatus—Experiments in Proximity', the author created a series of responsive video artworks collectively known as *Experiments in Proximity* which represent his most recent research as a notable media artist concerned with what he calls 'meta meta-cognition'. These simply designed and inventive artworks are central to the author's practice-research project, which falls under the more general rubric of embodied cognition and, as such, these works can be considered as 'enactive machines'.

These enactive machines, in their specific contexts and concerns, explore, for the author, the important concepts of structural coupling, enaction and sensorimotor contingencies, ideas that typically emanate from wide-ranging fields, from philosophy to cognitive science to media arts. The two most critical effects for Tonkin are that ideas allow him to create mechanisms for thinking about thinking (cognitive science), and, most significantly, for thinking about thinking about thinking as he has impressively accomplished in his own artworks. Ultimately, Tonkin believes that invention takes place in a non-linear fashion that highlights a co-emergence of theory and praxis grounded in the actual physical process of making.

The chapter by Brad Buckley and John Conomos, 'The Artist-run Initiative: An Agent That Blurs the Studio, Laboratory and Exhibition Space, Creating a Site for Inventiveness', underlines how invention within these unique artists' exhibition spaces, is integrally unique to contemporary art and how because of their unregulated nature and concerns are not saddled with the crushing ascendancy of corporate managerialism that is normative in the established global art world. Their historical origins emanate from the collectives of artists in the 1960s that were interested in working with experimental ideas and forms and operating in fringe spaces outside the mainstream gallery and museum system.

Essentially, artist-run initiatives (ARIs), such as in the accompanying case study by Alex Gawronski, a founding member of the Institute of Contemporary Art Newtown (ICAN), unequivocally demonstrate how inventive they are as an expression of aesthetic cosmopolitanism and experimental hybridism and as public spheres of artistic creativity and knowledge production that effectively critique the characteristic ideological aspects of Fordism, neo-liberalism and the culture industry as Enlightenment as mass deception.[15]

Buckley and Conomos maintain that ARIs are innovative, questioning and reflexive spaces of post-Fordist production and they insist on the deconstruction of the state-sponsored audit culture of the visual and creative arts, media institutions and critical reception.

Gawronski's informative case study, 'ICAN: Reinventing the Autonomy of the Artist-run Initiative', shows how in 2007 it was first instigated as a project by the author himself in association with fellow artists Carla Cescon and Scott Donovan, and also initially by the late Sydney artist Stephen Birch (1961–2007), who was a critical force behind the project. What Gawronski highlights in his carefully considered analysis of ICAN's importance as an ARI is how its very inventiveness as an experimental space is predicated on its evolving project to become an autonomous critical laboratory. As such ICAN over the past few years has been concerned with generating an ethos in Sydney where artists and academics in their respective shared histories and networks are mutually involved with 'reframing contemporary art as a visual form of non-instrumental *thinking*'. Further, another innovative attribute of ICAN has been its focus on presenting the works of already established and mid-career artists rather than the obligatory emerging artists who tend to be more representative of such spaces.

In essence, Gawronski clearly shows how ICAN has always in their multi-tiered approaches to contemporary art been committed to the continuing questioning of the neo-liberal commercialisation of the art world and its attendant culture of high spectacle. Salient to its wider climate of inventiveness as an ARI is ICAN's dedication to politically independent creative activity where knowledge is constantly foregrounded in non-commodity terms.

Chris Smith's chapter 'Fit to Burst: Bodies, Organs and Complex Corporealities' is an insightful, elaborate examination of Dagmar Reinhardt's and Lian Loke's 2012 imaginative and speculative work *Black Spring* which was exhibited at the University's Tin Sheds Gallery (a venue that has played an enormous role, historically speaking, in advocating contemporary art in Australia). Smith's equally speculative analysis of Reinhardt and Loke's work focuses on a Deleuzean approach to the seminal architectural question of thinking through the formation and life of space that is quintessentially central to how architects and artists work creatively.

The bold inventiveness of Reinhardt and Loke's interactive *Black Spring* resides precisely not in being traditionally an installation as such but rather best regarded—conceptually and performatively—as a sensate machine. This means that the work's multiple dancing bodies, contexts, processes and ideas are primarily concerned with uniting, in Smith's words, 'the corporeality of the sensate and the incorporeality of the machine'. The work involves bodies dancing, responding with each other, constantly changing through an infinite, complex structure of processes that happen inside and outside of the bodies themselves.

Reinhardt and Loke's work implicitly concerns itself with Deleuzean ideas of 'becoming', 'machine' and 'sensation' in its effective multifaceted project to transcend a disciplinary boundary of architecture by exteriorising depth from it, coercing forces from its inside.

Dagmar Reinhardt and Lian Loke's engaging contribution 'Entangled: Complex Bodies and Sensate Machines', is enlightening for its lucid exposition of their original design research activities as embodied in their three interactive spatial installations *The Black Project: Black Spring* and *Black Shroud*, collectively entitled *The Black Project* and *GOLD (Monstrous Geographies)*. All three works can be seen to be laboratories built over a design, construction and experiential period situated within an exhibition context and through the constant exposure of an unversed audience. All three installations critically represent an interdisciplinary conversational arena constituting a matrix of relationships fluctuating between several interlocking conceptual clusters of intriguing ideas of interdisciplinary design processes as re-invention, complex corporealities, inexactitude and excess, and sensate machines.

For Reinhardt and Loke, design research is a verb in that it incorporates speculative creative acts developing dialogue, doing, multi-direction and related processes of investigation that valorise alternative, devious and sideways results. Thus, their intriguing, fertile and expressive installations cogently demonstrate the authors' design research philosophy to represent explorations spanning 'body, space, code, and materiality in excess of architectural practice's privileging of form, function, economics'. Invention for them is constitutive of their creative and scholarly endeavours in the very act of unveiling important narratives, conceptual and semantic layers, and plateaus within a given potential work bridging architecture, choreography, engineering, interaction design and programming.

Sean Lowry's conceptually nimble and informative chapter 'Inventions Are Networks: Fostering the Liminal Play of Ideas' discusses how inventiveness is based on pre-existing networks of different contextual relations. Lowry persuasively makes the case for how the collaborative spirit of human activity favours the pliable capacity to recombine and repurpose existing ideas and objects in new unpredictable and challenging ways. This is especially the situation, Lowry argues, with the presiding ethos of our digital age. Hence, global networks are liminally enhancing and re-choreographing tensions between different individual and collective contexts and interests that underline cultural change, thereby fuelling invention.

The author provides numerous telling insights into how, through the Internet, it is critical to acknowledge that invention takes place when the lateral connectivity of digital social networks in association with the cross-disciplinary potential of open source digital databases directly produce new fields of innovation. He also contends that the broader the network of collective consideration that a creative work enters into, the more unlikely it is to be deemed critical in terms of its interdisciplinary significance.

Inventiveness, in its broadest sense, is therefore multi-scaled and dynamically responsive at micro, meso, and macro levels. The case study of Sydney composer Ivan Zavada that follows is a perfect instance of Lowry's argument that networked musical collaborations, across space and time, like Zavada's work, dramatically redefine relationships between performer and music, producing new forms of musical expression. The challenge is, according to Lowry, that one can't predict how new technologies and new markets will create invention, but it remains for all of us to locate new ways of fostering free, open global forms of experimentation.

Zavada, therefore, in his accessible analysis of his own recent network musical collaborations 'Expanding Sonic Space: An Antipodean Approach to Telematic Music' demonstrates how boundaries of conventional performance and creative practice in our society can be extended given our present mutating techno-culture. Composition is often regarded as an individual activity, but with the internet and new communication methods, Zavada shows us how the creative potential for a given composer transforms his role to one of being more of a coordinator of musical events over space and time. Depending on the network technology used, this will facilitate the coordination of different real-time musical events or pre-composed musical pieces being simultaneously performed in totally different places.

In such key works as *Jasmine* and *Antipode*, undertaken at the Sydney Conservatorium of Music, numerous participants in different locations (China, Canada, Australia) were engaged in different modes of interactivity and networked collaborations creating entirely new models of musical creativity. By amalgamating existing electronic music and different audiovisual processing techniques to increase the aesthetic concerns and boundaries of network musical performance, sonic space itself is creatively expanded.

Leading artist and media researcher Bill Seaman and equally renowned chaos physicist Otto E. Rössler, following their recent cutting-edge book *Neosentience* (2011) which espouses a new branch of scientific inquiry into artificial intelligence, have contributed the chapter 'Inventions and Recombinant Poetry' which appropriately enough closes the book. Their book chapter, which can be considered as an 'after-poem' to their book, is a highly inventive, playful and interdisciplinary exploration of creativity, complexity, connectivity, mind maps, robotics, memory, how we interface and interpret the world, and what it means to be human in our present globalised world.

It is a speculative, free-wheeling, poetic meditation on invention in the context of complexities of the arts and the sciences that ranges across many different disciplines such as psychology, computing, physics, cybernetics, logic, genetics, evolution, literature, art, architecture, neuroscience and linguistics. In its sprawling collage structure and trans-disciplinary concerns, theirs is a most vital and fitting 'book-end' chapter in that it encapsulates in its own inventive mode of writing the dynamic porous character of invention itself. In a critical sense, Seaman and Rössler's invaluable contribution is reminiscent of composer John Cage's mesostic poetry.

Think of 'inventor' or 'invention' and the images that come to mind might include Thomas Edison and Johannes Gutenberg, the polymer banknote or Wi-Fi, but probably not Buckminster Fuller and Picasso, the Torrens Title and the method of collage. All of these examples are inventors and inventions, as are the contributors to this volume and their works.

This volume contributes to a burgeoning body of evidence that broadens the view on invention from its narrow base of scientific, legalistic and economic understandings. Theories of collective intelligence (Pierre Levy), hyperintelligence (Mark Pesce) and cognitive surplus (Clay Shirky) all describe the ways ideas and knowledge circulate amongst global networks and thereby galvanise invention. Invention is also retrofitting, modification and hacking. As science fiction writer William Gibson famously commented in his short story *Burning Chrome*, 'the street finds its own uses for things' as user inventiveness leads to new uses and further inventions. In his book *Where Good Ideas Come From*, Steven Johnson describes this process of adapting existing components and re-purposing them as 'Exaption'. The contributions to this volume produce a new language by which to declare the value of a mulitiplicity of inventive approaches.

ENDNOTES

1   Wesley M. Cohen and Daniel A. Levinthal, 'Innovation and Learning: The Two Faces of R&D', *The Economic Journal* 99, no. 397 (1989).

2   M. Marsel Mesulam, 'From Sensation to Cognition', *Brain* 121, no. 6 (1998).

3   John A. J. Gowlett, 'Artefacts of Apes, Humans, and Others: Towards Comparative Assessment and Analysis', *Journal of Human Evolution* 57, no. 4 (2009).

4   Ralph L. Holloway, Jr., 'Culture: A Human Domain', *Current Anthropology* 10, no. 4 (1969).

5   Lewis. H. Sarrett, 'Research and Invention', *Proceedings of the National Academy of Sciences* 80, no. 14 (1983).

6   Norbert Wiener, *Invention: The Care and Feeding of Ideas* (Cambridge: MIT Press, 1993), 6.

7   To read John Maeda's comments in full on STEAM, see www.risd.edu/About/STEM_to_STEAM/, retrieved on 13 July 2013.

8   Charles Percy Snow, *The Two Cultures* (London: Cambridge University Press, 1998).

9   Stuart Cunningham, *Hidden Innovation: Policy, Industry and the Creative Sector* (Brisbane: The University of Queensland Press, 2013).

10   Cunningham, *Hidden Innovation*, 12.

11   Stefan Collini, introduction to *The Two Cultures* (London: Cambridge University Press, 1998), iv.

12   Dieter Daniels and Barbara U. Schmidt, introduction to *Artists as Inventors, Inventors as Artists*, ed. Dieter Daniels and Barbara U. Schmidt (Ostfildern: Hatje Cantz, 2008).

13   Daniels and Schmidt, *Artists as Inventors*, 10.

14   George Steiner, *Grammars of Creation* (New Haven: Yale University Press, 2002).

15   Max Horkheimer and Theodor W. Adorno. 1972. *Dialectic of Enlightenment*. (London: Allen Lane, 1944).

1

INVENTION AT THE LOCAL SCALE [CAPACITIES]

DISCOURSES OF INTERVENTION:
A LANGUAGE OF INVENTION

ANDY DONG

INVENTING CULTURAL MACHINES

PETRA GEMEINBOECK
ROB SAUNDERS

DISCOURSES OF INTERVENTION: A LANGUAGE OF INVENTION

ANDY DONG

I'd like to start with a classic psychological test of creativity. Imagine for a moment that you have been transported to another planet far from Earth. There, you encounter new life forms. Now, create a mental image of the life forms you encountered and sketch them.

If you sketched life forms similar to those produced by the majority of participants in the original experiment conducted by Ward and colleagues,[1] you probably associated the category of animals with alien life forms. From the conceptual structure of an animal, you then envisioned an exemplar of an animal such as a dog or a bird or perhaps even an alien animal from a science fiction movie. Then, since animals generally have a head, your alien life form probably has a head and more likely than not only one spherical head at the top of its body. Life forms generally also have a means for detecting light (i.e., vision) so your alien life form probably has some type of eyes on its head, and, again, likely two eyes. Even when participants were explicitly told not to produce animal-like creatures,[2] there was a general tendency for the alien life forms to conform to conceptual expectations of life forms (on Earth). In other words, we generally find it difficult to break out of existing conceptual structures even when required to generate new concepts. New ideas, more frequently than not, resemble old ones.[3]

In the field of design studies, this problem of being 'stuck' in existing conceptual structures resulting in the carryover of irrelevant or inappropriate elements from prior experience is known as the problem of design fixation.[4] Design fixation is a robust phenomenon;[5] its effects have been identified in a number of creative problem-solving situations[6] including architectural design,[7] industrial design[8] and engineering design.[9]

Like our fixations to prior conceptual structures, we can also become fixated to the organising principles of the disciplines within which we practice. So-called art–science or art–technology collaborations between artists and scientists and engineers have become a prominent mode of practice in contemporary arts. The 'movement' includes an ArtScience Manifesto.[10] Artists represented in this book who engage in this mode of artistic collaborative practice include Bill Seaman,[11] Mari Velonaki[12] and Petra Gemeinboeck.[13] Inventive collaborations between disparate disciplines, such as the arts and sciences, require individuals to break out of the fixation effects of prior experience, to behave as 'both and neither':[14] behaving as if the individuals were in/from both disciplines and also as if the individuals were from neither discipline, instead operating in a hybrid discipline determined by the exigencies of the creative practice.

While the benefits of the collaboration are often highlighted,[15] few artists describe the practical and intellectual challenges associated with collaboration across disciplines. As described by Mari Velonki in her collaboration with roboticists David Rye, Steve Scheding, and Stefan Williams, 'Wide cultural gaps can exist between disciplines, and are perhaps inevitable because of differences in acculturation and education.'[16] That is, intellectual conflicts in art-science collaborations are a reflection of disciplinary claims to knowledge. Fixation to disciplinary claims, whether explicit or tacit, is an obstacle to interdisciplinary collaboration because it both traps individuals in their prior experiences and in the conceptual structures and methods of knowledge acquisition of their respective disciplines. How then, in the inventive process between and across two or more disciplines, can individuals break out of their respective conceptual structures? How can new concepts be neither a simple expansion of existing conceptual structures nor a duplex copy of them—taking two disciplines and projecting them onto one sheet of paper, but one side of the paper is for one discipline, and the other side for the other?

In this essay, I explore how disciplinary collaborators can rupture their conceptual structures through the principles of 'discourses of intervention'. In psychotherapy, people who are socially meaningful to the person bring an intervention upon a person undergoing self-destructive behaviour. In architecture, an intervention is brought upon a space. The intervention rejects bonds to the history of space and place and, often, the norms of architecture. In collaboration, an 'intervention' is brought upon fields of knowledge to unmap and destabilise conceptual structures such that notions of specialised methods, canonical definitions, and established identities constituting a field of knowledge become eschewed. In so doing, discourses of intervention open up a conceptual space for the invention of new ideas.

I begin by exploring how a key element of collaboration, the transmission of knowledge between individuals, can create obstacles to inventive collaboration using Basil Bernstein's concept of the pedagogic device.[17] Bernstein's concepts of classification and framing and hierarchical and horizontal knowledge structures are employed to examine the problems of legitimacy in inter-disciplinary knowledge sharing and creation. I then develop the principles of discourses of intervention for art–technology collaborations. Drawing upon examples of artistic and technical concepts from the collaboration between Petra Gemeinboeck and Rob Saunders and between Mari Velonaki and David Rye described in Section 2, I illustrate how traversing the hierarchical and horizontal knowledge structures involves shifts in meaning in both directions. The shifts can produce a space from which new inventive concepts can emerge.

BACKGROUND

Basil Bernstein was a residential family caseworker in East London before he became an educational researcher and scholar.[18] It was during this time that he noticed differences in the ways that the middle class, the aspirational working class, and the working class communicated amongst one another and between each other. The field of sociolinguistics was still in its infancy and so the concept that language would be mediated by social relations had yet to be fully theorised.

Addressing this problem, Basil Bernstein's body of research dealt with the transmission of knowledge and its entwinement with economic and social class and the economic rationalisation and capitalisation of knowledge. For Bernstein, it was not just that the surface-level features of communication, such as word choice, differed between speakers owing to their socioeconomic class; it was that the underlying communication codes differed.[19] He developed a code theory to explain how 'Curriculum defines what counts as valid knowledge, pedagogy defines what counts as valid transmission of knowledge, and evaluation defines what counts as a valid realisation of the knowledge on the part of the taught.'[20] The two principal codes in his code theory are classification and framing. Classification refers to the strength of the boundary of a discipline and framing refers to the strength of control over the selection, organisation, pacing and timing of knowledge transmission.[21] Communication by individuals within disciplines reflects these underlying codes. Disciplines could be more porous such that the boundaries between disciplines are weaker (weaker classification). Curriculum and the structure of classroom teaching could strictly regulate the content and sequence of content (stronger framing).

Bernstein is careful to note, though, that his code theory is intended to apply beyond the school context to any context in which relationships exist for the purpose of knowledge production–reproduction. One of the most socially progressive aspects of his theorising deals with pedagogic rights. Bernstein proposes three pedagogic rights: 1) individual enhancement; 2) social, intellectual, cultural and personal inclusion; and 3) participation in a discourse and practice that must have outcomes.[22] What is remarkable about these rights is that they set the pre-conditions for the creation of new intellectual possibilities. The transmission of knowledge can do more than replicate an image of the knowledge reproduced onto a new subject when the underlying codes governing the transmission shift from stronger to weaker classification and from stronger to weaker framing.

Bernstein describes this form of knowledge transmission, the 'mystery of the subject', as that which allows for the potential for possibilities to be thought:

> By the ultimate mystery of the subject, I mean its potential for creating new realities. It is also the case, and this is important, that the ultimate mystery of the subject is not coherence, but incoherence: not order but disorder, not the known but the unknown.[23]

It is in Bernstein's later work on the forms of knowledge, including their underlying organising principles and the social basis for differentiation that he developed the concepts of horizontal and vertical discourse.[24] Bernstein describes horizontal discourse as common, everyday knowledge that operates in segmented contexts. The forms of horizontal discourse are 'likely to be oral, local, context dependent and specific, tacit, multi-layered, and contradictory across but not within contexts'.[25] Stated simply, horizontal discourse is 'everyday' or 'common' knowledge obtained through general practice. In contrast, vertical discourse is a systematically and hierarchically structured set of knowledge that manifests as specialised language and criteria for both the production and the circulation of the knowledge. Within vertical discourses, Bernstein further distinguishes between hierarchical knowledge structures, in which theory is the metric of achievement, and horizontal knowledge structures, in which specialised languages and the perspectives and positions that they produce constitute the currency of legitimacy. Hierarchical knowledge structures consist of a set of general theories or propositions, which organise and integrate knowledge at lower levels of hierarchy. The natural sciences are characterised by hierarchical knowledge structures, with general theories such as the three laws of thermodynamics at the top of the hierarchy. Fields such as the arts or philosophy have a horizontal knowledge structure since they have specialised forms of criticism specific to contexts and historical periods such as Byzantine or modern art and specialised forms of inquiry such as phenomenology or logic. One consequence of their different forms of knowledge structures is that knowledge in these fields advances in different ways. The natural sciences progress as new and more general and integrative theories displace and subsume existing theories. The displacement of theories does not make sense for horizontal knowledge structures because specialised forms of criticism relevant to Byzantine art are not necessarily relevant to modern or contemporary art. Instead, fields having horizontal knowledge structures advance through the introduction of a new language: 'A new language offers the possibility of a fresh perspective, a new set of questions, a new set of connections, and an apparently new problematic, and most importantly, a new set of speakers.'[26]

To sum up, Bernstein's code theory introduces a set of concepts, classification and framing, and hierarchical and horizontal knowledge structures, to explain the transmission of knowledge between individuals. Bernstein's focus on communication is essential in the context of art–technology collaboration and invention. Communication between collaborators in this context is not simply about sharing ideas and knowledge, problem-solving and negotiation and other interpersonal pragmatics.[27] Communication is implicated in enacting their collaborative creative practice. Their communication performatively enacts and produces that which it names.[28] By engaging in a conversation with the object they are shaping, their creative work is materialised through these conversations. The communication intertwines the collaborators in representing that which can be conversed and said. The communication is a particular way of creating meaning and of representing a projected reality through language. Thinking of cross-disciplinary art–technology collaboration and invention as performative opens up the space of terminology to characterise the collaboration as one which operates partially through a linguistic system of representation to: 1) aggregate—to blend ideas and concepts; 2) accumulate—to scaffold ideas and concepts; and 3) appraise—to evaluate and assess ideas and concepts; and through these enact the collaboration and make representable and recognisable the creative work.[29] To illustrate the value of Bernstein's concepts on understanding and facilitating art–technology collaboration, I will explore a solution to cross-disciplinary knowledge production by proposing a set of principles for communication in art–technology collaborations: principles of discourses of intervention.

PRINCIPLES OF DISCOURSES OF INTERVENTION

Bernstein's concepts of classification and framing and hierarchical and horizontal knowledge structures contain several features useful to the understanding of the aspects of disciplines to which practitioners can be fixated and strategies for unfixing them. Practitioners will bring to any art–technology collaboration a specific form of disciplinary communication. According to Bernstein's theory, communication by individuals will reflect the underlying codes of the individual's discipline. As shown in a number of sociology of education studies, these codes are often transparent to the individuals[30] and the individuals may not recognise what type of communication is required by the other discipline. Inadvertently, there may be struggles over which code dominates, what some researchers call a 'code clash'.[31] A code clash occurs when the code(s) underlying communication differs between individuals. Code clashes generally lead to the sort of problems reported in cross-disciplinary collaboration such as cultural gaps between disciplines.[32] Bernstein's concepts offer potential solutions to establish 'a shared language that incorporates elements of each collaborator's area of expertise'[33] and more generally enables the theorisation of knowledge practices associated with cross-disciplinary art–technology collaboration and invention.

Existing accounts of cross-disciplinary art–technology collaboration and invention ask that practitioners share 'basic ethical and ideological positions'.[34] Starting from this proposition, in the rest of the paper, I draw upon Bernstein's concepts to develop principles for a 'discourse of intervention' as 'positions'. A discourse of intervention is brought between individuals when the underlying codes for their communication clash due to their disciplinary specialisations. It is deployed because the nature of collaboration between practitioners is communicative and discursive and produces a new discourse— that is, a new body of knowledge about and for their creative work. Communication between the collaborators must progress from conveying meaning to producing meaning.

The first problem in producing meaning, though, is overcoming 'code clash' of permissible boundaries of knowledge established by disciplinary specialisations and the collaborators' accumulation of canonical facts or narratives. As Bernstein theorised, fields control the transmission of knowledge, both within the field and to/from outside the field, through stronger or weaker forms of classification and framing. This creates a barrier to knowledge production across art-technology collaborations. The reality for the collaborators is that the new knowledge they produce must nonetheless be able to withstand the legitimacy of the discipline of each collaborating partner.[35] The new knowledge that the collaborators produce must at once be structured by their respective fields of knowledge as well as structuring the structure of the new fields of knowledge they can produce. While it is true that their new knowledge is a social construction, it is also true that the constructed knowledge is shaped by the beliefs of the individuals regarding how legitimate knowledge should be constructed and thereby included into the disciplinary canon.[36] To open up the space for new knowledge requires a commitment to weakening classification and framing.

PRINCIPLE 1: Discourses of intervention weaken
the boundaries of disciplines.

PRINCIPLE 2: Discourses of intervention weaken control
over the permitted content of knowledge.

The second problem is one of integrating hierarchical and horizontal knowledge structures—in other words, recontextualising specialised knowledge so as to make it accessible to others outside the discipline. I suggest that the recontextualisation requires a form of linguistic technicalisation: taking a discipline-specific use of a concept and using it in an analogous manner in the other discipline. Discourses of intervention establish ontogological connections between concepts in disciplines having hierarchical knowledge structures with analogous concepts in disciplines having horizontal knowledge structures and vice-versa. The diagram shown in Figure 1 depicts examples of this recontextualisation. Terms on the upper vertical axes are taken from a discipline having a hierarchical knowledge structure in this example, (probability theory) whereas terms in the lower vertical axes are derived from Judith Butler's concept of performativity,[37] a specialised language (horizontal knowledge structure) for the interrogation of gender. The practice of connecting any two pairs of concepts, such as *'a priori'* and 'system of meaning', from disciplines having opposite knowledge structures requires the insertion of the concept of *'a priori'* into the specialised language of analysis of Butlerian performativity. *A priori* probability is the probability of an event without any prior information. In the absence of prior information, an *a priori* estimate of a probability is based on assumptions about the characteristics of the system. Similarly, a system of meaning precedes gender performance, which produces the ontology of a recognisable gender performance. Gender exists *a priori* of the gender performance.

Likewise, terms on the horizontal axes represent concepts that draw upon knowledge developed within vertical discourses but their use in a collaborative context is likely to be context dependent and specific and their meanings will have been developed through experience and enculturation. These concepts will likely be deployed toward the achievement of specific, immediate and highly relevant goals to the collaboration. In other words, they have properties of horizontal discourse. The danger, though, is that the discourse may remain mired in the local context. Discourses of interventions transform horizontal discourses into a new specialised language for modes of questioning and production of knowledge relevant to the creative work. They develop a new horizontal structure of knowledge arising from the art–technology collaboration through the introduction of a new language with a new set of questions and a new set of voices. The new 'construct' that is measured and argued from a specific 'position' is neither a simple expansion of existing conceptual structures nor a duplex copy of them; rather, it adds to the accumulation of knowledge across specific contexts.

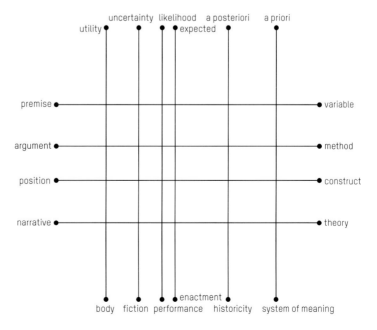

Figure 1.1: Recontextualisation of knowledge across vertical and horizontal knowledge structures. Figure adapted from Piet Mondrian, *Composition with Red, Yellow and Blue* (1921).

CONCLUSIONS

Discourses of intervention operate in three ways:

They weaken boundaries of disciplines.

They weaken control over permissible knowledge.

They introduce horizontal knowledge structures by transforming their local, context dependent and specific knowledge, which is grounded in hierarchical knowledge structures.

Bernstein characterises the creation of a new 'language' as sometimes necessary to 'challenge the hegemony and legitimacy of more senior speakers'[38] and to mark a new territory. In art–technology collaboration and invention, principles of discourses of intervention contribute to demapping hierarchical knowledge structures in order to reveal new possibilities from them. The intervention reads the process of requiring hierarchical and horizontal discourses to be acquired for the collaboration not as a process of remapping, hybridising or genericising, but rather a process of crossing lines of division and separation.

The notion of discourses of intervention expresses both a mode of working and an effect of working through the aims of the collaborative creative work. Ideally, discourses of intervention function to effect collaboration that creates new specialised languages. Through communication assisted by discourses of intervention, the de-fixation from disciplinary conceptual structures comes to inhabit the creative work.

ACKNOWLEDGEMENTS

This paper originally appeared in the *Proceedings of New Constellations: Art, Science and Society* as 'Discourses of Intervention: A Language for Arts Collaboration' by Petra Geimenboeck and Andy Dong. The author gratefully acknowledges permission of the Museum of Contemporary Art Australia to revise, extend, and publish the new version. The conference was part of an Australian Research Council–Australia Council for the Arts Linkage Project. The author formerly collaborated with Petra Geimenboeck on her series *Impossible Geographies*.

ENDNOTES

1   Thomas. B. Ward, 'Structured Imagination: The Role of Category Structure in Exemplar Generation', *Cognitive Psychology* 27, no. 1 (1994).

2   Thomas B. Ward and Cynthia M. Sifonis, 'Task Demands and Generative Thinking: What Changes and What Remains the Same?', *The Journal of Creative Behavior* 31, no. 4 (1997).

3   Thomas B. Ward et al., 'The Role of Graded Category Structure in Imaginative Thought', *Memory & Cognition* 30, no. 2 (2002).

4   David G. Jansson and Steven M. Smith, 'Design Fixation', *Design Studies* 12, no. 1 (1991); A. Terry Purcell and John S. Gero, 'Design and Other Types of Fixation', *Design Studies* 17, no. 4 (1996).

5   Steven M. Smith and Julie Linsey, 'A Three-Pronged Approach for Overcoming Design Fixation', *The Journal of Creative Behavior* 45, no. 2 (2011).

6   Carlos Cardoso and Petra Badke-Schaub, 'Fixation or Inspiration: Creative Problem Solving in Design', *The Journal of Creative Behavior*.

7   Purcell and Gero, 'Design and Other Types of Fixation.'

8   Carlos Cardoso and Petra Badke-Schaub, 'The Influence of Different Pictorial Representations During Idea Generation', *The Journal of Creative Behavior* 45, no. 2 (2011).

9   Vimal Viswanathan and Julie Linsey, 'Examining Design Fixation in Engineering Idea Generation: The Role of Example Modality', *International Journal of Design Creativity and Innovation* 1, no. 2 (2013); Robert J. Youmans, 'The Effects of Physical Prototyping and Group Work on the Reduction of Design Fixation', *Design Studies* 32, no. 2 (2011).

10 Bob Root-Bernstein et al., 'ArtScience: Integrative Collaboration to Create a Sustainable Future', *Leonardo* 44, no. 3 (2011).

11 Bill Seaman, 'The Hybrid Invention Generator', *Leonardo* 38, no. 4 (2005); 'Oulipo | Vs | Recombinant Poetics', *Leonardo* 34, no. 5 (2001).

12 Mari Velonaki et al., 'Shared Spaces: Media Art, Computing, and Robotics', *Computers in Entertainment* 6, no. 4 (2008); M. Velonaki et al., 'Fish-Bird: Cross-Disciplinary Collaboration', *MultiMedia, IEEE* 15, no. 1 (2008).

13 Petra Gemeinboeck and Rob Saunders, 'Urban Fictions: A Critical Reflection on Locative Art and Performative Geographies', *Digital Creativity* 22, no. 3 (2011); Petra Gemeinboeck et al., 'On Track: A Slippery Mechanic-Robotic Performance', *Leonardo* 43, no. 5 (2010).

14 Ruth West et al., 'Both and Neither: In Silico V1.0, Ecce Homology', *Leonardo* 38, no. 4 (2005).

15 Dana Plautz, 'New Ideas Emerge When Collaboration Occurs', *Leonardo*.

16 Mari Velonaki et al., 'Sharing Spaces: Risk, Reward and Pragmatism', in *New Constellations: Art, Science and Society* (Sydney, NSW, Australia: Museum of Contemporary Art, 2006), 76.

17 Basil Bernstein, *Pedagogy, Symbolic Control, and Identity: Theory, Research, Critique*, ed. Donaldo Macedo, Revised edn, Critical Perspectives Series (Oxford: Rowman & Littlefield, 2000).

18 Alan R. Sadovnik, 'Basil Bernstein (1924–2000)', *Prospects: The Quarterly Review of Comparative Education* XXXI, no. 4 (2001).

19 Basil Bernstein, *Class, Codes and Control*, 4 vols., vol. 1—Theoretical Studies towards a Sociology of Language (London: Routledge, 2003).

20 *Class, Codes and Control*, vol. 2—Applied Studies Toward a Sociology of Language (London: Routledge & Kegan Paul, 1975), 85.

21 *Pedagogy, Symbolic Control, and Identity: Theory, Research, Critique; Class, Codes and Control*, 1—Theoretical Studies towards a Sociology of Language.

22 *Pedagogy, Symbolic Control, and Identity: Theory, Research, Critique*.

23 *Class, Codes and Control*, vol. 1—Theoretical Studies towards a Sociology of Language.

24 'Vertical and Horizontal Discourse: An Essay', *British Journal of Sociology of Education* 20, no. 2 (1999).

25 'Vertical and Horizontal Discourse: An Essay', 159.

26 'Vertical and Horizontal Discourse: An Essay', 163.

27 Robert B. Arundale, 'Conceptualizing 'Interaction' in Interpersonal Pragmatics: Implications for Understanding and Research', *Journal of Pragmatics*, in press.

28 Andy Dong, 'The Enactment of Design through Language', *Design Studies* 28, no. 1 (2007).

29 Dong, 'The Enactment of Design through Language.'

30 Lucila Carvalho, Andy Dong, and Karl Maton, 'Legitimating Design: A Sociology of Knowledge Account of the Field', *Design Studies* 30, no. 5 (2009); Rainbow Tsai-Hung Chen, Karl Maton, and Sue Bennett, 'Absenting Discipline: Constructivist Approaches in Online Learning', in *Disciplinarity: Functional Linguistic and Sociological Perspectives*, ed. Frances Christie and Karl Maton (London: Continuum, 2011); Alexandra Lamont and Karl Maton, 'Choosing Music: Exploratory Studies into the Low Uptake of Music GCSE', *British Journal of Music Education* 25, no. 3 (2008); Karl Maton, 'The Wrong Kind of Knower: Education, Expansion and the Epistemic Device', in *Reading Bernstein, Researching Bernstein*, ed. Johan Muller, Brian Davies, and Ana Morais (London: Routledge, 2004).

31 Alexandra Lamont and Karl Maton, 'Choosing Music: Exploratory Studies into the Low Uptake of Music Gcse; Karl Maton, 'Knowledge-Knower Structures in Intellectual and Educational Fields', in *Language, Knowledge and Pedagogy: Functional Linguistic and Sociological Perspectives*, ed. Frances Christie and James R. Martin (London: Continuum, 2007).

32 Velonaki et al., 'Sharing Spaces: Risk, Reward and Pragmatism.'

33 West et al., 'Both and Neither: In Silico V1.0, Ecce Homology,' 287

34 Velonaki et al., 'Fish-Bird: Cross-Disciplinary Collaboration,' 12

35 Velonaki et al., 'Fish-Bird: Cross-Disciplinary Collaboration', 12.

36 Bernstein, *Pedagogy, Symbolic Control, and Identity: Theory, Research, Critique*.

37 Judith Butler, *Bodies That Matter: On the Discursive Limits of 'Sex'* (New York: Routledge, 1993).

38 Bernstein, 'Vertical and Horizontal Discourse: An Essay.'

INVENTING CULTURAL MACHINES

PETRA GEMEINBOECK
ROB SAUNDERS

Human history is intertwined with the history of machines. According to some scholars, it is the drive to create machines that makes us unique,[1] and many agree that human evolution began with the invention of tools.[2] It seems safe to say that we can think of human evolution only as a co-evolution of humans and machines. They allow us to control and even create our environment, have empowered us to dominate the world, and have become our mirror image in which to study ourselves. This co-evolution is the starting point, context and experimental playground of our collaborative practice in art and technology. Rather than extending the debate on 'The Singularity' and whether machines will eventually dominate their human inventors, we're interested in the intermingling of the two, and the continuum between what we consider our human life and that of our machines. In a way, our source of inspiration is human inventiveness itself, and our motivation is to invent a reality for the two—humans and their creations—to encounter one another.

Our artwork *Zwischenräume* (In-between Spaces, 2010–12) develops from an unusual concoction of walls, curious robots, and surveillance technology. The work nestles itself into the human environment by embedding robots into the architectural skin of a gallery, sandwiched between the existing wall and a temporary wall, which resembles it. Each machine agent is equipped with a motorised hammer or punch, a camera and a microphone to interact with its environment and communicate with the other machines. Programmed to be curious about their world, the robotic agents are intrinsically motivated to explore, and, once they have ripped open the wall, they begin to study its human inhabitants. Doing so, they physically manifest and inscribe their perception and communication processes into our built environment. The wall becomes the milieu through which the machines express their desires, leaving behind marks and open wounds.

INVENTING A MACHINIC ASSEMBLAGE

The starting point of our interdisciplinary collaboration was our common interest in the open-ended potential of creative machines autonomously acting within the human environment. From the computational creativity researcher's point of view, the embodied nature of the agents allows for situating and studying the creative process within a complex material context. For the artist, this collaborative practice opens up the affective potential to materially intervene into our familiar, human-created environment, bringing about a strange force, seemingly with an agenda and beyond our control.

One unique aspect of *Zwischenräume* is that the robots are hidden, at least at first, behind—what audiences believe to be—an existing wall. We are interested in deploying robotics as a medium of intervention to shift the focus from representation to performativity. As such, the robots become part of an affective assemblage entangling human agents and nonhuman agents. Following Broeckmann's definition of the machinic as an aesthetic principle, we were interested in 'process rather than object, with dynamics rather than finality, with instability rather than permanence, with communication rather than representation, with action and with play'.[3] This unstable context of invention has challenged our experimental process. We needed to invent embodied agents such that we could: control to the extent that they remained operative, while behaving and adapting autonomously in an ever-changing environment; develop a configuration that couples these machine agents with other agents—walls—in ways in which the wall could become an active agent that the machines need to adapt to, while at the same time being transformed and structurally compromised by this material 'play'; and devise a setting that would allow us to inject this dynamic process into our existing, familiar environment. While we have discussed the inner workings of *Zwischenräume* in more detail in other writings,[4] here we give preference to the constitutive contextual layers of our inventive process. It is perhaps unnecessary to say that this process wasn't as clear-cut and linear as the above sequence of challenges would suggest. On the contrary, in contrast to technological inventions, the drivers for artistic inventiveness are primarily problem finding, rather than problem solving. Thus, while the above challenges could be seen as both conceptual and technical problems to solve, more often than not, found or identified 'problems' simply open up another set of possible paths, each unfolding a series of discoveries and unforeseen points of departure.

We can imagine that this account of our process of discovery resonates with many scientists. What is fundamentally different is that we didn't start out with a hypothesis, and we didn't have a rigorous set of protocols, neither to adhere to, nor at our disposal. Rather, our experimental process could be described as a kind of material thinking, at points driven by our conceptual desires, and at other times driven by the materials we encountered and assemblages that emerged from the process. In the process, concept (our ideas and motivations for bringing these material actors together) and matter (the material agents and their capacities to interact with other agents) mingle to a degree where they become inseparable. Perhaps another significant distinguishing aspect between an artistically driven inventive process and many other creative inventive processes is that there is no optimal solution. Even if there was, this isn't the goal; rather, it is a process of allowing both the 'goal' and the 'solution' to emerge from a series of iterative conceptual–material negotiations. This requires both collaborators to sometimes abandon what they know, and to allow unexpected or emergent confluences of material behaviours and, for example, programmed behaviours to set the course.

Figure 1.2: *Zwischenräume*, 2012, installation view, GoMA, Queensland.
Photo courtesy of the artists.

Figure 1.3: *Accomplice*, 2013, installation view, Artspace, Sydney. Photo: silversalt photography.

The machinic assemblage as aesthetic principle merges with our motivation to also situate the work in our respective academic research fields: experimental media arts and computational creativity. In both our research practices, we are interested in developing an ontology of performative capacities. Andrew Pickering talks about performative capacities as 'a decentered perspective that is concerned with agency—doing things in the world—and with the emergent interplay of human and material agency'.[5] To provide an example, it is vital that *Zwischenräume*'s machine agents have the capacity to adapt autonomously and act proactively so that both the material and social encounter happen without our prescribing it a course to unfold. As the agents' embodiment evolves based on their interaction with the environment, the robots' creative agency affects processes out of which agency itself is emergent.

Gordon Pask's *Musicolour* (1953), processing the sound from musical instruments to control a light show, is a pioneering work with regards to placing human and machine performance in a level playing field setting. The work did not simply convert sounds in a pre-scripted manner; it adapted its performance in response to how it 'perceived' the musicians' performance. If its inputs from their performance became too continuous, *Musicolour* changed the frequency ranges or rhythms it listened to and responded only when it registered those.[6] As this affected the musicians' improvisation, the machine became a co-performer.

In applying the computational model of curious agents,[7] we empowered *Zwischenräume*'s robots to become self-motivated and proactive agents that, once embodied and embedded in our human environment, nestle themselves into our social fabric, even if only temporarily. This is where the encounter happens. The machines are driven to learn by discovering novel elements in their environment, even if this means that they themselves introduce difference by knocking at the wall until it gives way and their 'world' can offer a new, interesting element to study. From the audiences' point of view, the machines' eager engagement and the impact it has on *their* environment brings with it an alien, uncanny force, able to unsettle our sense of control and security. It is not as simple as a dualist encounter, however; machines have never been and will never be a force from the outside. The complicit entanglement gets further uncovered once the machines have opened up their environment and the robots begin to discover the world on the other side of the wall (see Figure 1.2). They not only respond to the audiences' presence and behaviours, but also have the capacity to perceive the audience with a curious disposition. As the audience peeks into the holes in an attempt to understand what drives these machines, the machines look back at them. The newly discovered human inhabitants bring an insatiable amount of novelty to their world. The audience is thus implicitly implicated in the work's unfolding affair between the robots and their environment, and becomes complicit, rather than being invited to control the course of events. This produces an interesting shift, where the work not only performs for the audience's eyes, but audience members also perform for the work's voyeuristic gaze.

Machines that act on what could be thought of as a 'level playing field' with humans not only open up the cultural potential of creative human-machine collaborations, but also the political potential of human–nonhuman assemblages that unhinge the human's central and superior position. Pickering romantically describes it as 'the dance of human and nonhuman agency', a dance that involves 'indefinitely open-ended searches of spaces of agency'.[8] This is also the dance of material thinking, where nonhuman, material performativity becomes an arena of negotiation that expands, diverts and, literally, messes with our doing and thinking. In our experimental practice, we constantly struggle with material agencies,

Figure 1.4: *Zwischenräume*, 2010, installation view, MuseumsQuartier 21, Vienna.
Photo courtesy of the artists.
Figure 1.5: *Zwischenräume*, 2010, installation (view from outside), Forum Stadtpark, Graz, Austria.
Photo courtesy of the artists.
Figure 1.6: *Zwischenräume*, 2010, installation (view from behind the wall), MuseumsQuartier 21, Vienna.
Photo courtesy of the artists.

in creative ways, including those that we often assume to be passive and static. The material character of the plasterboard wall in *Zwischenräume*, for example, makes it an active player, whose organic frills (i.e., fragments of gypsum board) not only visually transform the machines' actions but also actively affect their perceptions and behaviours.

Coming back to the process of invention, we argue that it is not only the many voices and knowledges from the inventors' cultural and social context that play a significant part, but also the many nonhuman agencies, material and machinic, that are constitutive of the inventive process. Often these agencies only emerge and play themselves out in Pickering's 'dance of human and nonhuman agency'. According to Karen Barad, agency always is an enactment, 'not something that someone or something has'.[9] In the process of making, eventually the indefinitely open-ended number of searches shrinks to a limited number, constrained by available resources but also the need for the process to eventually gel into one, coherent performance. Importantly, however, our performative environments, never present a final, fixed result, but rather assemble a certain stage in the process or configuration of actors that remains open and dynamic. Thus, embracing the unfixed and unknown in the principle of assemblage shapes both the authors' iterative, experimental process of developing the work and the audience's experiential process as the configuration of players unfolds.

THE ARTWORK AS A CULTURAL MACHINE

Artistic inventiveness does not only concern the production of an original 'artefact', nor is it the outcome of the experimental technical development in our work. We see it as a strategic intervention into cultural and social discourse, where the artwork is its aesthetic materialisation and the technology is at once the means of production and the complicated, investigated protagonist. For *Zwischenräume*, assembling itself at the edge of both the human and the nonhuman realm, this conundrum is a source of inspiration itself. *Accomplice* (2013), our latest machine-augmented environment, extends the notion of social agency into the machine realm: the machine inhabitants of our built environment become social agents and the wall becomes a shared territory for them to connect, cooperate and conspire.

As machines become part of our social fabric, the co-evolution of humans and machines is complicated by the increasing agency and autonomy of our machines and our growing intimacy with them. And this is perhaps the first layer of encounter that our works evoke. Yet *Zwischenräume* and *Accomplice* also challenge the confining human–machine dualism by letting the encounter of humans and machines unfold in an unexpected, perhaps slightly precarious scenario that aims to allude to a broader notion of 'machine'. If we had been only interested in creating machines that autonomously destruct walls, we would have placed them inside the gallery space to perform for the audience in plain sight. *Zwischenräume* and *Accomplice* instantiate cultural machines that, as they unfold and become, reach beyond the conceptual bounds of the technological apparatus, and operate as an abstract machine that installs itself transversally to all heterogeneous machine levels, the material, cognitive, affective and social.[10] Promoting a politics of change, the artistic inventive act operates like a rupture, disrupting the familiar, habitual assemblage to create a new affective assemblage.[11] Hence, we argue that the employment of robotics as a medium of intervention to create an affective assemblage that radicalises the human–machine relationship is a significant inventive aspect of our collaborative practice.

Looked at in more pragmatic terms, many parts of this assemblage cannot be bought off-the-shelf, and the act of assembling involves a great deal of experimentation and invention. We do not invent technologies to facilitate the making of the work or to (re)present an experience, but rather the technological invention becomes the artefact itself, or perhaps, more precisely, enacts the performance. Simon Penny calls it a 'machine-artwork', where 'the embodied affective experience is integrally related to and cannot be separated from the material manifestation of the artefact'.[12]

Shannon's *Ultimate Machine* is perhaps also the ultimate machine-artwork. A simple wooden case with the size and shape of a cigar box, the only thing we can see on the outside is a switch. Once switched, one can hear motors buzzing, the lid opens, and a hand emerges—only to turn the switch off. The hand retreats back inside the box, the lid shuts and the machine is an unassuming box again. After Arthur C. Clarke's haunting encounter with the *Ultimate Machine*, he wrote: 'There is something unspeakably sinister about a machine that does nothing—absolutely nothing—except switch itself off'.[13] The machine's sole purpose is to subvert the very notion of what a machine is, and, thus, extends what a machine can be.

COLLABORATIVE INVENTIONS IN ART AND TECHNOLOGY

Invention and creation are at the heart of both artistic and engineering processes, albeit creation often has a too unwieldy undertone to be used to describe the engineer's motivation, and invention often mingles too closely with innovation and its corporate baggage to be embraced by artists. Media art, and in particular the area that is often referred to as Art and Technology, has a history of artists and engineers collaborating or individuals who embrace the skills and values of both cultures. In contrast to the engineering domain, inventing an optimised solution to an already known problem rarely drives an experimental artistic process. Rather, as Jill Bennett succinctly puts it, 'experimentality manifests as a disposition, a drive to question, transgress and reinvent that in turn inflects the particular exploratory processes or 'methods' of art making'.[14] To the artist, the world does not pose a spectrum of problems that need to be solved or a messiness that needs to be modelled and controlled. This defiance of rigorously formalised and reproducible approaches renders art a more inviting and promising field 'to pursue eccentrically motivated research'.[15]

Much of what separates art and engineering practices is rooted in their institutionalisations and the complex social meshwork they produce, and, arguably, 'art is apt to exceed any institutional designation, confounding expectations about what it is and where it belongs'.[16] As in any collaboration, in the end, however, it is people with their personal interests, motivations and backgrounds who meet, engage and work together. This very personal aspect cannot be underestimated, and, inter- or transdisciplinary collaborations are better thought of as collaborations between people coming from different disciplinary backgrounds or driven by different motivations, informed by their disciplines, rather than collaborations between disciplines.

Our collaboration presents a rare constellation of an artist and an engineer, co-creating artworks, that acknowledges both as equal authors of the work. Each of us is involved in the problem-finding and -solving in both arenas: the conceptual–aesthetic and the technical–agential. Despite the long history of artists and engineers collaborating, such horizontal co-authorship still seems radical. This is where the collaborative practice falls apart into its disciplines and institutions again, and only the artist can claim authorship of the artwork. More often, both cultures see the engineers' contribution as a service to facilitate the artists' work, rather than as part of a mutual collaboration that transforms each other's work and has the potential to flow back into their respective fields. While, naturally, each collaboration is different, it seems strange that even in the field of experimental art and technology, where the machine-artwork emerges from an intimate mingling of conceptual and technical problem-finding and -solving, the engineer is seen as only enabling the making of the artwork but not its co-creator.

Let's briefly take a closer look at this by going back to an important point in the history of Art and Technology. Experiments in Arts and Technology (EAT), founded in 1966 by the engineers Billy Klüver and Fred Waldhauer and artists Robert Rauschenberg and Robert Whitman, was seen as 'a tentacular service organisation, whose mission was to facilitate the work of artists in any field or discipline'.[17] All artworks that came out of EAT have been associated only with the artists, e.g. Robert Rauschenberg's *Solstice* (1968), which he developed in collaboration with the engineer Robby Robinson. Billy Klüver provides us with a lively, intimate account of their inventive collaborative process: in February 1968, Rauschenberg called Robinson about a new work for the 'Documenta' in Kassel, which opened at the end of June, 1968. In their first meeting about this new work, according to Robinson, Rauschenberg 'wanted to try to achieve something that would involve people with his work but not in such a sophisticated way that they were mystified by how it happened, but so they could really see how it happened'.[18] Klüver continues,

> As they talked about what the possibilities were, Robinson told Rauschenberg about going through several sets of sliding doors at Schiphol Airport in Amsterdam to get into the main terminal and that he thought they could put something on the doors to alert you to the next one you would encounter. A few days later, according to Robinson, Rauschenberg called and said: 'Robby, I have an idea.' ... So I went there and he had a whole plan worked out in his mind for *Solstice*, a series of sliding doors mounted on a platform. The light would be constant and the only movement would be the mechanical movement of the doors. He explained that he was thinking of an image that you could view either from the front or from the rear. In addition, you could actually walk through the layers of the image and find how the total image was derived from the various [coloured] overlays.[19]

Robinson's role was to create the physical installation and its pneumatically driven sliding Plexiglas doors, while Rauschenberg composed the screens to became the layers of the image for people to walk through.[20] This opens up the question: does the artwork come into being by placing Rauschenberg's screens onto the industrially manufactured sliding doors, or is it the environment or experience this installation creates by allowing people to move through the layers of an image? We think it's the latter, and Robinson's contribution, starting with the familiar experience of the airport's set of sliding doors, cannot be reduced to the level of a facilitating service. In addition, it is mainly due to Billy Klüver's passionate accounts that we can engage in these pioneering events in art and technology today. Klüver was interested in finding 'new means of expressions for artists ... and to find out where they stood in relation to the society that was sending men to the moon'.[21] But he didn't believe that this was a one-directional transformation. In an interview with Edward Shanken, Klüver said: 'I believed that the artists would influence the engineers and then change the technology. Of course, the point is that the artists would work with engineers and change the engineers.'[22]

Isn't it this visionary belief, or naïve conviction, that often frames the starting point compelling both artists and researchers to walk down yet unknown paths? We must acknowledge here that collaborations in art and technology are still happening only at the fringe of the art world. Even if accepted, the interactions, transformations, and inventions that can only happen in this specific coming together of people with different disciplinary backgrounds are often not recognised as constitutive of the artistic process, only as enabling.[23] Yet, does engineering value the changes that the engineers may bring back to the field? All too often, the field of engineering is blind to the cultural context that opens up new paths. We need to understand how much of the 'problem solving' happens in the very act of the coming together of different cultural forces. A 'solution' may not be a singular event or an optimised outcome, but rather a new window into the complex interactions and entanglements that every invention is situated within.

ENDNOTES

1   Bruce Mazlish, *The Fourth Discontinuity: The Co-Evolution of Humans and Machines* (New Haven: Yale University Press, 1993), 216.

2   Marie Landau, *Narratives of Human Evolution* (New Haven: Yale University Press, 1993), 51. Rodney Brooks, Flesh and Machines (New York, NY: Pantheon Books, 2002), 5.

3   Andreas Broeckmann, 'Remove the Controls. Machine Aesthetics', 1997. Accessed May 20, 2013. v2.nl/events/machine-aesthetics.

4   Please see: 'Other Ways of Knowing: Embodied Investigations of the Unstable, Slippery and Incomplete', *Fibreculture Journal* 18, 2011, 9–34. 'Creative Machine Performance: Computational Creativity and Robotic Art' in Proceedings of the Fourth International Conference on Computational Creativity (ICCC 2013), University of Sydney, 12–14 June 2013.

5   Andrew Pickering, 'Cybernetics and the Mangle: Ashby, Beer and Pask', *Social Studies of Science* 32:2 (2002), 414.

6   Usman Haque, 'The Architectural Relevance of Gordon Pask', *Architectural Design* 77:4 (2007), 56.

7   Rob Saunders and John Gero, 'A curious design agent: A computational model of novelty-seeking behaviour in design'. (Paper presented at the Sixth Conference on Computer Aided Architectural Design Research in Asia. CAADRIA, 2001).

8   Pickering 'Cybernetics and the Mangle: Ashby, Beer and Pask', 414.

9   Karen Barad, *Meeting the Universe Halfway: Quantum Physics and the Entanglement of Matter and Meaning* (Durham, NC: Duke University Press, 2007), 178.

10 Felix Guattari, *Chaosmosis* (Bloomington and Indianapolis: Indiana University Press, 1995), 35.

11 Simon O'Sullivan, 'From Aesthetics to the Abstract Machine: Deleuze, Guattari, and Contemporary Art Practice' in *Deleuze and Contemporary Art*, ed. Simon O'Sullivan et al. (Edinburgh: Edinburgh University Press. 2010), 199.

12 Simon Penny, 'Bridging Two Cultures: Toward an Interdisciplinary History of the Artist-Inventor and the Machine-Artwork' in *Artists as Inventors—Inventors as Artists*, ed. Dieter Daniels et al. pa(Ostfildern, Germany: Hatje Cantz Verlag, 2002), 149.

13 Arthur C. Clarke, *Voice Across the Sea* (New York: HarperCollins, 1958), 159

14 Jill Bennett, 'What is Experimental Art?' *Studies in Material Thinking*, 8 (2012): 1.

15 Penny, 'Bridging Two Cultures: Toward an Interdisciplinary History of the Artist-Inventor and the Machine-Artwork', 155.

16 Bennett, 'What is Experimental Art?', 1.

17 Sylvie Lacerte, '9 Evenings and Experiments in Art and Technology: A Gap to Fill In Art History's Recent Chronicles' in *Artists as Inventors—Inventors as Artists*, ed. Dieter Daniels et al. (Ostfildern, Germany: Hatje Cantz Verlag, 2002), 161.

18 Robby Robinson (1979) quoted in Billy Klüver and Julie Martin, 'Working with Rauschenberg' in *Robert Rauschenberg: A Retrospective*, ed. W. Hopps et al. (New York: Guggenheim Museum Publications, 1998), 320.

19 Robinson, quoted in Klüver and Martin, 320.

20 Klüver and Martin, 'Working With Rauschenberg'.

21 Harriet De Long, '9 Evenings: Theatre and Engineering' (manuscript); Series 1. Project files; box 2, '9 Evenings', 1966; Experiments in Art and Technology; Records, 1966–1993; Getty Research Institute, Research Library, Accession no. 940003.

22 Edward Shanken, 'I Believed in the Art World as the Only Serious World That Existed', Interview with Billy Klüver, in *Artists as Inventors—Inventors as Artists*, ed. Dieter Daniels et al., 176.

23 There are a small number of academic institutions, centres and study programs that are committed to interdisciplinary research in the area of art and technology and that promote the potential of mutual transformations and an emerging Third Culture. Without providing a comprehensive list here, we would like to point to the MIT Media Lab, promoting a unique, anti-disciplinary culture; the Art | Sci Center + Lab at UCLA, California, and its vision of artists working in the lab and scientists working in the art studio; the Ars Electronica Centre's pioneering exhibition program that seamlessly engages audiences in artistic, engineering and scientific inventions and Simon Penny's interdisciplinary graduate program in Arts, Computation and Engineering at the University of California, Irvine.

INVENTION AT THE
LOCAL
SCALE
[SENSIBILITIES]

THE 'CHARACTER' AND THE 'ALGORITHM':
AN ESSAY ON TECHNOLOGY AND ART

DAN LOVALLO

ART AND ROBOTICS —
A BRIEF ACCOUNT OF ELEVEN YEARS
OF CROSS-DISCIPLINARY INVENTION

MARI VELONAKI
DAVID RYE

Figure 2.1: *Eucalyptus salmonophloia murray darling seed 2*, 2013, The University of Sydney.
Copyright Dan Lovallo. Photo: Karl Connolly.

THE 'CHARACTER' AND THE 'ALGORITHM': AN ESSAY ON TECHNOLOGY AND ART

DAN LOVALLO

This essay will explore technology and art. The topic area is wide so I will have to necessarily limit my scope in two ways: first, to the influence of algorithms in technology-mediated art, and second, to four pieces recently completed by University of Sydney artists and researchers. The main focus will be on two new robotic installations that debuted at Artspace Visual Arts Centre in Sydney in May 2013. The first piece is by the internationally renowned artist Mari Velonaki entitled *The Woman and the Snowman*. The second piece is *Accomplice* by Petra Gemeinboeck and Rob Saunders. The final pieces are by the author. The first of these is *Eucalyptus salmonophloia murray darling*, also produced in May 2013. It is included in order to discuss the burgeoning importance of additive manufacturing (also known as 3D printing) in art and society at large. The final piece, *Constellation 3—Exploded Juius*, is an installation made with an algorithm but constructed with simple rather than complex materials.

I begin with *The Woman and the Snowman*. This installation, which filled a space approximately 20 metres by 20 metres, consists of two video projections—one of female android robot Repliee Q2 by Professor Hiroshi Ishiguro from Osaka University. She is dressed in a fine red gown, a tall wispy tree behind her, and snow covers the ground. The other projection is a snowman in his natural setting—snow. The next piece of the composition is another robot but this robot is not anthropomorphic. In fact a better term for it would be kinetic sculpture, as it moves to a score devised explicitly for the installation. Every so often, the kinetic sculpture changes speeds (four possible) and rotation (two) while keeping time with the haunting multi-layered score by Nikolas Doukas. The kinetic sculpture moves at a rate several magnitudes faster than the movement in the projections. There are a few screens that rotate inside the kinetic sculpture. One of the screens is of a winter scene, another of a 1970s Greek television show. All of these images are from the artist's past. The sculpture was developed at the University of Sydney's Australian Centre for Field Robotics where Velonaki works with her long-time collaborator David Rye and others in the Centre for Social Robotics.

The trees move imperceptibly around the Snowman as the sun moves across the sky. At a similar pace the Woman's hand slowly turns as she powers down.

The overall impression of the installation is that one has landed in a strange but benign foreign world. Nothing is threatening; in fact everything is fragile: we see the Woman at her most vulnerable, as she is powering down. The Snowman obviously has inherent vulnerabilities, and the use of nostalgic pictures in the kinetic sculpture imbues the finely crafted sculpture with an unexpected softness.

The resonances within the piece are many and subtle. It turns out the character of the Snowman is the most 'true' to himself. He is not trying to be anything other than a snowman whereas the Woman is an android portraying a woman. While the Snowman has some obvious frailties, he is much more robust than the Woman who would be ruined if she became wet and cold. Then, there is a time duality between the projections and the kinetic sculpture and a score that ties everything together. The algorithms control every movement of the robots and mix the worlds of human and machine, expanding the audience's consciousness of machines beyond the thesis of a (mathematically) functional relation between human brains and machines in Norbert Wiener's *Cybernetics* into an environment wherein humans and machines become synchronized. This synchronization is not unlike our synchronization of behaviours with other humans, such as yawning when others yawn, or walking at the same pace as others walk. The algorithms underlying the *The Woman and the Snowman*, while functionally deployed to control the movement of the robots, also function as a social interface to attend the audience to the patterns of movements and sounds created by the robots in their interaction with each other. The time shift forces the participants to attend to one piece at a time and then at a distance to the whole composition. As I watched participants, I noticed that the kinetic sculpture moved people around the room. They would come back to it and then go away. It not only psychologically tied the pieces together in Velonaki's mind but it physically created a touchstone for the participants, which I found surprising. The result is a sublime contemporary masterpiece. Only someone with her diverse influences—performance art, kinetic sculpture, robotics, Japanese cinema, the Theatre of the Apparatus and poetry, among others— could have pulled it off. Finally, it is worth saying she must be a master collaborator in addition to her other gifts in order to engage numerous highly skilled colleagues.

Petra Gemeinboeck and Rob Saunders (henceforth robococo) began their collaboration in 2007. Gemeinboeck's background is in interactive media art. Her practice crosses the fields of architecture, interactive installation, mobile media, robotics, and visual culture. Saunders' primary research interest is the development of computational models of creativity. The development of computational models of creative processes provides opportunities for developing a better understanding of human creativity, producing tools that support human creativity and possibly creating autonomous systems capable of creative activity.

In this show, these artists collaborated on *Accomplice*. *Accomplice* is a group of four robots—two on two out of four walls in a 20 metre by 20 metre space— who can communicate with each other both audibly and visually. Obviously, they can also communicate with the participants using the same senses, but the humans and the robots do not necessarily understand the communications in the same way.

The robots communicate by signalling their presence by banging on a wall with an implement that can poke a hole through the plasterboard. The robots signal back by repeating an identical pattern or a similar pattern. Gemeinboeck reflects that the signalling is inspired by birdsong. This is like no bird I have ever heard before because it is loud and jarring. All four robots communicate to the three others but the others do not always respond. When they do, once again, it is loud. One feels a sense of foreboding, as if the building might fall at your feet. I would not be surprised if this were a future goal. A large part of the real goal is to see what behaviours arise as the walls come apart. They walls fall apart and the artwork emerges much like other emergent systems—there is unpredictability but also regularity. The piece relates closely in spirit to Jean Tinguely's, *Homage to New York* (1960) but the robots function more effectively.

The piece is essentially a performance by robots. However, the robots' behaviours are not readily apparent. Watching the robots have their way with the walls is like watching a new animal one has never seen. It takes time to infer what they are up to. The social bonding right now involves (sometimes) replicating patterns. In the future the goal is to both model and signal curiosity on the part of the robots. This is a rich paradigm that will undoubtedly be useful both in art and science. So it is not surprising that a major interest of robococo is the co-evolution of man and machine. Partially because of this interest, they explicitly do not use humanoid shaped robots because they believe this will be but a small fraction of the robots in use. Witness that the first commercially available robotic 'maids' such as the iRobot Roomba® bear no resemblance to a humanoid. Through experimentation with these robots, they test and imagine the future.

One interesting aspect of these pieces is the role of algorithms in controlling the behaviour of the robots, specifically their interaction with each other and their environment but not necessarily with the audience directly. In fact, the robotics in *Complicit* may or may not pay much attention to the gesticulations of humans peering through the holes produced by the robots to catch the robots' attention. Robococo's work is breaking new ground on algorithms that will allow heretofore human characteristics such as curiosity to be embedded in robots. Thus far in this essay, I have not disambiguated the algorithm from the robotics. All robots more or less need complicated algorithms. At one extreme Saunders is breaking new ground on modelling creativity algorithmically. Typically, creative computational synthesis algorithms depended either upon (human) pre-defined metrics of creativity, such as an absolute measure of novelty, or upon human feedback to determine if the auto-generated actions by the algorithm should be deemed creative by a human observer. In Saunders' model, creativity is a function of both an algorithm's individual evaluations of creativity combined with emergent social definitions of creativity. That is, the robots individually define what is creative based upon their own set of interactions with the environment, and thereby influence emergent social definitions of creativity because the other robots are implicated in the production of the environment. Algorithmically, the robots use a novelty detector, which is based on the Wundt curve. If the stimulus is 'too' novel, then it is considered strange rather than interesting. Likewise, if the stimulus is too similar to prior stimuli, then it is considered mundane. Within a certain bandwidth of difference, the stimulus is considered novel. The algorithm causes the robot to prefer stimuli that are similar-yet-different to stimuli experienced previously.[1]

On the other extreme, the algorithm is the means of form generation and topological control. In one recent piece, *Eucalyptus salmonophloia murray darling* (May 2013), I used a branching algorithm to combine the branching of a specific tree and river basin. Subsequently, I printed the result out using additive manufacturing. Whilst I programmed the algorithm, the algorithm was in control of both the eventual form and the mode of production of the form through additive manufacturing. I am neither the first artist nor the last to make use of algorithms. 'Traditional' art is liable to explode as additive manufacturing machine prices plummet. In fact, Boeing is planning on 'printing' planes.

Algorithms can even be used to organise work that is made with very simple materials. For example, a work like Tara Donovan's *Haze* (2003), which is made through a laborious repetitive process, has kinship to algorithmic work in that the piece takes shape through repeated routine manipulations. In *Constellation 3—Exploded Julius* (2013), I used an algorithm to organise simple materials like bottles, wire and coloured water into complex shapes that would have been nearly impossible without the algorithm.

The conceptual artist Sol LeWitt perhaps anticipated the coming of the computer age when he first wrote instructions for his wall drawings. At the time, critics enamored with the notion of the heroic artist questioned whether someone giving simple instructions for painting a wall but not physically painting the wall himself was actually an artist. The question of human agency becomes more critical as algorithms become more complex and play a larger role in our lives. A couple of examples will illustrate the issues that arise. For example, there is one level of complication if an algorithm, perhaps with a random

component, makes an artwork. In other words, if the author of the algorithm is uncertain as to the outcomes that will be obtained, to what extent can the author be considered the creator of the output? The question of agency is further complicated if the algorithm is meant to create further algorithms or takes in data from its operating environment as the basis for its actions, an environment that cannot be predicted in advance by the algorithm's authors. At what point does the algorithm take on a life of its own? Is there a point at which, in a generative sense, the author of the original algorithm can no longer take credit or receive blame for the subsequent inventive output?

These questions come into sharper relief when we move from the realm of art into the financial world where algorithms already drive a large amount of high-frequency trading on exchanges like the New York Stock Exchange. In this realm, several relatively small (compared to the size of the market) companies (most recently Knight Capital on 1 August, 2012) have caused either large parts of the market or the entire market to shut down based on "rogue" algorithms. These algorithms are not simply finding stocks to trade and directing orders. They are, much like the robots in *The Woman and the Snowman* and *Accomplice*, producing an environment of their own making: a marketplace of electronic trades. Like the actions of the robots in *Accomplice*, if the automated trading robots make a mistake, or make a hole in the wall 'in the wrong place', it is no longer possible for the system to go back (without considerable human intervention). And, like the robots in *The Woman and the Snowman* or in Velonaki's earlier work The *Fish-Bird* Series, they can continue to perform in their marketplace 'oblivious' to the presence of humans or not. Clearly, only the algorithms implicated in high frequency trading are 'oblivious' to any calamities that they may have created. Clearly, when an algorithm written and 'owned' by a relatively small company can cause nearly incalculable financial damage, it is time to revisit the laws written for an earlier age. Both patent and liability law need updating in light of the algorithms that fly our planes, trade our retirement accounts and, in some cases, make our art. It is in these respects that algorithms in art raise a political issue: perhaps we no longer live in the anthropocene but rather in the 'algorscene', an age when algorithms define and control political and social ecosystems. Beyond their aesthetic ambitions, these works of art become quite useful for thinking about what our inventions may themselves invent.

ENDNOTES

1   Saunders, Rob. (2007). 'Towards a computational model of creative societies using curious design agents', in Gregory O'Hare, Alessandro Ricci, Michel O'Grady, and O. Dikenelli (eds.), *Engineering Societies in the Agents World VII* (Berlin: Springer-Verlag, 2007), 340–353.

ART AND ROBOTICS —A BRIEF ACCOUNT OF ELEVEN YEARS OF CROSS-DISCIPLINARY INVENTION

MARI VELONAKI
DAVID RYE

Our collaboration began in 2002, when Mari Velonaki visited the Australian Centre for Field Robotics (ACFR) at the University of Sydney to discuss possible collaboration on her new project, *Fish-Bird*. Although Velonaki had extensively utilised technology in her work, *Fish-Bird* was a very ambitious project as it required the design and construction of two autonomous robots—thus the need for high-level collaboration with roboticists. The project was enthusiastically received by roboticists David Rye, Steve Scheding and Stefan Williams. The team was drawn together through a shared passion in the area of human–robot interaction. Another important starting point for us was our common interest in the creation of novel human–machine interfaces.

From Velonaki's point of view, the interest was in the creation of haptic interfaces that promote explorations of new models of interaction between the participant and an interactive kinetic object, such as a robot. The opportunity to place a robot in a public space such as an art gallery or museum was both attractive and challenging to Rye, Scheding and Williams. Gallery spaces are open to the general public, and attract a wide variety of visitors, from children to the elderly, so in a technologically driven interactive artwork the human–machine interface must be intuitive and robust to unexpected events. This is a significant challenge to robotics.

Early in our collaboration we began to converse, exchange ideas and investigate mechanisms for cross-disciplinary research funding, as well as work on our first project *Embracement*. One motivation for creating *Embracement* was to determine if the team could work together, and to find the most effective ways of collaborating.

Embracement is a light-reactive installation. The work explores perception and interpretation of interpersonal interaction, through the device of optical illusion. Within a custom-made photo-dynamic crystal screen, fleeting image sequences of two women appear. All sequences commence with the women standing facing each other, and with the younger woman striding towards the elder one. Subsequent interactions between the characters are placed ambiguously between affection and violent rejection. The nature of the relationship remains unresolved to the viewer as the projected image is interrupted by a shutter. The viewer is left with ephemeral glowing red afterimages of the women. This transient memory of the screen is due to the physics of the crystals, which absorb light and re-emit it at red wavelengths. This is not a software (e.g., Photoshop) effect, but a consequence of physical photoluminescence. Projection of images under computer control allows for indefinite variation in the repeating image sequences.

An important starting point for us was our common interest in the creation of human–machine interfaces that are effective, intuitive, and fulfilling. Working on projects that will be displayed in museums and art galleries is very different to working in the lab or in the 'field' where only trained operators use computer-controlled machinery. This aspect is important to roboticists: art spaces are open to the general public and the human–machine interfaces used in such spaces cannot rely on a comprehensive knowledge of the inner workings of the system—they must be intuitive and robust to unexpected events. Furthermore, the average three-month duration of a museum exhibition poses a significant challenge to robotics practitioners. Once installed, an interactive robotic artwork has to operate unattended for the duration of the exhibition, operated and maintained by museum staff who often do not have technical backgrounds. During this time the artwork will be visited by thousands of people. These untrained members of the general public may find themselves interacting with a robotic system, which must function reliably and behave in engaging ways. From the beginning of our collaboration we have emphasised the importance not only of the interface between interactants and the artwork, but also the simplicity of the operational interface between the installation and the gallery staff. Achieving sufficiently high reliability to operate successfully under these constraints is a significant challenge to contemporary robotics science.

The early stages of our collaboration required extensive discussions to clarify and establish a shared language that could bridge dissimilarities in our fields, and to develop our ideas of what extended cross-disciplinary collaboration means and what it should accomplish. These discussions led us to agree that the following three characteristics are essential for a successful cross-disciplinary collaboration. Firstly, collaborating individuals must have both shared and individual goals. It is important for a group to identify through discourse a clear, common goal and to commit as a group to attaining that goal. Within the group, participants should have individual goals that are not necessarily shared by the whole group. Secondly, trust is essential for an extended, fully engaged collaboration—what Mamykina et al.[1] term a 'full partnership'. An exchange of skills alone is not sufficient to keep people working together; compatibility of personality and agreement on basic ethical and ideological positions are essential prerequisites for building trusting relationships that can survive the challenges associated with the development of both new concepts and new technologies. Thirdly, the development of a shared language is essential to cross-disciplinary collaboration. Wide cultural gaps can exist between disciplines, and are perhaps inevitable because of the differences in acculturation and education. The cultural gap between the creative arts and robotics is pronounced, although there are similarities between the two disciplines in their grounding in experimental practice. A shared language is essential in building a shared vision and trusting relationships within the partnership.

Our group's definition of collaboration is deliberately broad, but is also altruistic: collaboration requires that participants interact as equals, respecting each other's strengths and experiences, and should explicitly acknowledge the contributions of all involved. In shared decision-making, there should be no need for any party to compromise. In a fully engaged collaboration, there needs to be space—a generosity of spirit—for discussion, accommodation and contribution from various points of view. The process of accommodation is quite different from compromise. Development of trust can lead to fruitful relationships within a team and space for accommodation without the need for personal or professional compromise.

Furthermore, we argue that the ultimate success or failure of truly collaborative projects, such as those described here, is best measured by the scholarship of the project outcomes. We define 'scholarship' in terms of: knowledge of 'best practice' in one's own discipline; the advancement of best practice; and the dissemination and uptake of the research outcomes by one's peers. Scholarship therefore also implies legitimacy within the discipline of each collaborating partner.

In the following sections we discuss the creation of new robotic forms and other responsive art projects. What a robot could look like and how it could behave is discussed in relation to what has influenced aesthetic and behavioural decisions through four cross-disciplinary projects: *Fish-Bird, Fragile Balances, Diamandini* and *The Woman and the Snowman*.

FISH-BIRD

In 2003 our group (Velonaki, Rye, Scheding and Williams) received one of the first two Australian Research Council (ARC) Grants to be awarded for cross-disciplinary art and science research. The Australia Council for the Arts was a partner in the project through the Synapse initiative, together with Australian Network for Art and Technology, Artspace Visual Arts Centre, Museum of Contemporary Art, and Patrick Technology and Systems. This ARC/Synapse grant supported the creation of our first major project *Fish-Bird: Autonomous Interactions in a Media Arts Environment (2003–06)*, which led to the interactive robotic installation *Fish-Bird: Circle C—Movement B*.

Fish-Bird: Circle C—Movement B is an interactive installation that explores the dialogical possibilities between two autokinetic objects (two robotic wheelchairs) and their audience. Assisted by integrated miniature thermal printers, the chairs write intimate letters on the floor, impersonating two characters, Fish and Bird, who fall in love but cannot be together due to 'technical' difficulties. In their shared isolation, Fish and Bird communicate intimately with one another via movement and text.

Spectators entering the installation disturb the intimacy of the two objects yet create the strong potential for other dialogues to exist. The spectator can see the traces of their previous conversation on the floor, and may become aware of the disturbance that they have caused. Dialogue occurs kinetically through the wheelchairs' perception of the body language of the audience, and the audience's reaction to the unexpected disturbance would be to converse about trivial subjects, like the weather. Through emerging dialogue, the wheelchairs may become more 'comfortable' with their observers, and start to reveal intimacies on the floor again.

Each wheelchair writes in a distinctive cursive font that reflects its 'personality'. The written messages are subdivided into two categories: personal messages communicated between the two robots, and messages written by a robot to a human participant. Personal messages are selected from fragments of love-letters offered by friends, from the poetry of Anna Akhmatova,[2] and from text composed by Velonaki.

The wheelchair was chosen as the dominant object of the installation for several reasons. A wheelchair is the ultimate kinetic object, since it self-subverts its role as a static object by having wheels. At the same time, a wheelchair is an object that suggests interaction: movement of the wheelchair needs either the effort of the person who sits in it, or of the one who assists by pushing it. A wheelchair inevitably suggests the presence or the absence of a person. Furthermore, the wheelchair was chosen because of its relationship to the human—it is designed to frame and support the human body almost perfectly and to assist its user to achieve physical tasks that they may otherwise be unable to perform. In a similar manner, the *Fish-Bird* project utilises the wheelchairs as vehicles for communication between the two characters (Fish and Bird) and their visitors.

Many aspects of the system design are strongly influenced by the desire to conceal the underlying technological apparatus. It should not be obvious to a spectator/ participant how a wheelchair moves, promoting rapid engagement with the work and focusing attention on the form of interactive movement. As a consequence of this conceptual and ideological consideration, standard electrical wheelchairs could not form the basis of the autokinetic objects in the artwork. A robotic wheelchair together with all associated electronics and software were custom designed for the project.

The science and technology used to realise *Fish-Bird* is highly complex, and is in some areas at the limit of what can presently be achieved in robotics practice. As an engineered system, the installation contains at least eight computers: three or more for the vision and scanning-laser system that tracks the movement of people and wheelchairs in the installation space; one computer to control and coordinate robot motion and text; and two microcontrollers with Bluetooth connectivity in each wheelchair (two). The system is in effect a distributed data fusion network that combines information from vision, laser and infrared sensors to construct a real-time representation of what is happening in the installation space. The system attempts to classify the intent of participants on the basis of their movement, and then generates robot motion and text messages in response to the perceived intent.

To place *Fish-Bird* within a contemporary art context, we argue that it confronts major continuing issues and concerns regarding interaction through the human—machine interface. The dialogical approach taken in this project both requires and fosters notions of trust and shared intimacy. It is intended that the technology used in the project will be largely transparent to the audience. Going further than a willing suspension of disbelief, a lack of audience perception of the underlying technological apparatus will focus attention on the poetics and aesthetics of the artwork, and will promote a deeper psychological and/or experimental involvement of the participant/viewer.

Robots in the context of popular culture have historically been associated with anthropomorphic representations. Although they represent characters, the robots in *Fish-Bird* are neither anthropomorphic nor are they pet-like or 'cute'. The audience internalises the characters through observation of the words and movements that flow between the characters, and between the characters and the audience, in response to audience behaviour. Through movement and text the artwork creates the sense of a person, and allows an audience to experience that person through the perception of what is not present.

FRAGILE BALANCES

Velonaki, Rye and Scheding continued their collaboration in the second work of the *Fish-Bird Series—Circle D: Fragile Balances*. In this work, two luminous cube-like objects appear to be floating above the surface of a lacquered wooden structure that perches on impossibly slender legs. Each object is comprised of four crystal screens where 'handwritten' text appears, wrapping around it conveying a playful sense of rhythm. The text represents personal messages that flow between the virtual characters of Fish and Bird, and in that sense each object is a physical embodiment of a character. The objects can be lifted from their wooden stand and handled freely by participants. Handling provides an interface that facilitates bidirectional communication between the participants and the artwork in a playful way.

If a gallery visitor picks up one of the cube-like objects from its floating base, the text becomes disturbed and barely readable, influenced directly by the movement of the visitor's hands. The sensitive structure of the personalised messages flowing between the two fictional characters remains disturbed as long as the visitor moves or turns the object quickly or abruptly. The only way that the participant can allow the messages to again flow around the object is to handle it with care— gently and softly cradling the object in his/her hands in concert with the rhythm of the 'handwritten' messages. If visitors do not handle the luminous cube objects, the work stands on its own as a complete sculptural piece containing an internal kinetic element—the moving text.

In this installation the objects provide an interface that facilitates bidirectional communication between the participants and the artwork. *Fragile Balances* deals with concepts of fragility, trust and communication by playfully challenging the participants to pause and enter the rhythm of the floating words and the dialogues that they lead to.

Interaction occurs between artwork and audience through the reactive objects as information passes from the object to the participant and from the participant to object. Another linkage involves the latent relationship between the two participants—the objects become the medium for a participant to become aware of the existence of a virtual character. The moment that a participant chooses to pick up and hold one of the objects, s/he becomes an avatar for this character in the actual physical space of the installation. Should *both* participants then choose to vocalise their individually received fragmented messages as they appear around the surface of their objects, or to move close to each other in order to read each other's messages, then the dialogue between the virtual characters is manifested and completed in the physical space.

The text used in the works represents the everyday dramas that arise within the minutiae of life—a stream of eternal hopes, of forgotten promises, of love letters that are never received: the things that keep us alive, even in the most troubling times. To reach into this fragile stream of mysterious text, the audience must attain a moment of stillness.

The realisation of *Fragile Balances* required design and construction of small, precise wooden cubes that house the LCD screens, as well as containing and mounting the internal computers, battery packs and sensors. The cubes also had to remain closed and tamper-proof during exhibition while allowing for daily start-up and shutdown and wire-free battery recharging. Conventional wood-working technology was unable to manufacture the wooden housings with sufficient precision. The solution was to cut all wooden elements for the cubes from a slab of Black Bean timber using a computer-controlled milling machine. The five sides of the wooden frame were assembled using epoxy glue and internal stainless steel alignment pins; no screws were used. Stainless steel locating pins were also used to align the base of the wooden assembly as it is closed up to the other five sides by screwing a conical charging assembly into the base of the box. Once assembled it is not apparent that the cube can be opened. Since the cubes remain closed during an exhibition, a permanent magnet is used to trigger internal reed switches to start and stop the computers inside them.

Each cube contains two Linux computers, each controlling two LCD screens and coordinating the text displayed across all four screens. A microcontroller provides power management and interfaces to a triaxial accelerometer that is used to sense how a participant handles the object. Built-in Bluetooth wireless is used to communicate between the two objects and to an external computer that can provide a data connection to the *Fish-Bird* robots. Approximately half of the volume of a cube is filled with Lithium-polymer battery packs to give an extended run time before recharging is needed. The cubes are arguably the most complex hardware system built to date at the Australian Centre for Field Robotics.

Figure 2.3: *Fish-Bird: Circle C — Movement B*, 2005. Centre for Social Robotics, The University of Sydney. Copyright Mari Velonaki. Photo Paul Gosney.

Figure 2.4: *Circle D: Fragile Balances*, 2008. Centre for Social Robotics, The University of Sydney. Copyright Mari Velonaki. Photo Paul Gosney.

Figure 2.5: *Current State of Affairs*, 2010. Cockatoo Island, Sydney. Copyright Mari Velonaki. Photo Paul Gosney.

Figure 2.6: *The Woman and the Snowman*. 2013, installation view,
Artspace, Sydney. Photo: silversalt photography.

Figure 2.7: *The Woman and the Snowman*. 2013, installation view,
Artspace, Sydney. Photo: silversalt photography.

DIAMANDINI

Diamandini is a five-year ARC research project between Velonaki and Rye at the Centre for Social Robotics within the Australian Centre for Field Robotics, University of Sydney, and the Creative Robotics Lab at the National Institute for Experimental Arts/ COFA, University of New South Wales. The continuing project aims to investigate human–robot interaction in order to develop an understanding of the physicality that is possible and acceptable between a human and a robot within social spaces.

With *Diamandini,* Velonaki wanted to make a new robot that would take the experimentation begun with *Fish-Bird* further by adding the element of interaction via touch. It was important that the interaction be one-to-one: one human, one robot.

The Greek word for interactive transliterates as amphi-dromos (amphi: around on both sides of dromos: street or road). Thus, it is defined as a middle point where two roads meet. In English, the preposition 'inter' means 'between' or 'among'. Inter-action therefore signifies between or among actions—a meeting point beyond action and reaction and prior to discourse, a brief moment of recognition between two parties. In this meeting point of recognition and identification, Velonaki intends to use the moment as a stage to test if intimate human-to-human interactions can serve as an analogue for human-to-robot interactions.

The original intent of the *Diamandini* project was to create a robot that was non-representational and non-anthropomorphic. As Velonaki started experimenting with a variety of abstract sculptural forms, although interesting in shape and structure, as the artist/creator she found it extremely difficult to assign behaviours to them that could lead to emotional activation of the spectator/participant.

With Fish and Bird, although the wheelchair robots are certainly not anthropomorphic, it was inevitable for the participants to assign personalities to what was not there, since a wheelchair is a socially charged object that signifies the absence or the presence of a person. The dialogues expressed in written text between the two characters in *Fish-Bird* and the storyline further assisted the participants to feel a momentary connection to the Fish and Bird characters.

These considerations influenced Velonaki's decision to create a humanoid robot. This was a challenging decision, especially when deciding how the robot should look. Velonaki did not want *Diamandini* to have a typical humanoid robot aesthetic. After a long period of reflection, she began to think of *Diamandini* as a female sculpture. In her mind *Diamandini* had a diachronic face that spans between centuries, a style that could be reminiscent of post-World War II fashion influences and, at the same time, with futuristic undertones.

Figure 2.8: *Diamandini*, 2012. Medieval and Renaissance Gallery, Victoria and Albert Museum, London. Images courtesy of the Victoria and Albert Museum, London.

Diamandini is small—only 155 cm high. Velonaki wanted her figure to be small and slender so that people did not feel threatened by her when she 'floats' in the installation space. She wanted *Diamandini* to look youthful, but not like a child, and for her age not to be easily identifiable. According to Velonaki, 'In my mind she is between 20 to 35 years old. Because I am a woman I feel more comfortable working with a female rather than a male representation.'

The construction of *Diamandini* was a multi-stage process, involving a sculptured prototype terracotta head, a custom-tailored fabric dress made over a wooden armature, and high precision 3-dimensional laser scanning and manipulation of the scanned data, followed by computer-aided design modelling. The external shell was made using stereolithography, an additive manufacturing process that uses computer-controlled UV lasers to polymerise a resin in very thin layers.

As it was important for *Diamandini* to appear 'floating' across the floor of an installation space, a commercial motion base could not be used. An omnidirectional motion platform containing three computer-coordinated driven and steered wheels was designed and constructed. The wheels contain a clever kinematic mechanism invented by Mark Calleija of the ACFR that gives pure kinematic rolling motion regardless of the steering and driving motions of the wheels. The omnidirectional motion base decouples *Diamandini*'s facing direction and rotational motion from the direction and speed of her movement so that she can glide backwards, forwards or sideways, and transition smoothly between these movements. Current work in progress will add tilt to her body.

Diamandini's arms were designed with four rotational degrees of freedom each—two at the shoulder and two at the elbow— so that the articulation of her arms was simple, yet able to make the range of gestures required for interaction. Sensing, including distributed touch sensing, and actuation will be added to the current prototype arms.

The project has also involved considerable research into touch sensing and the transmission and interpretation of social messages and emotions via touch. David Silvera Tawil (now a Postdoctoral Fellow at the Creative Robotics Lab, NIEA) as a PhD student at ACFR developed techniques based on electrical impedance tomography (EIT) that can be used to implement flexible and stretchable artificial 'sensitive skin' to facilitate the interpretation of touch by robots during human–robot interaction. Silvera Tawil developed a classifier based on the 'LogitBoost' algorithm and used it to identify social messages—such as 'acceptance' and 'rejection'—and emotions transmitted by touch to an arm covered by the EIT-based skin. Experiments demonstrated, for the first time, that emotions and social messages present in human touch could be identified with accuracies comparable to those of human-to-human touch.[3] Future work in the project will transfer these techniques to *Diamandini*.

THE WOMAN AND THE SNOWMAN

The Woman and the Snowman investigates and questions the boundaries of what can be perceived as real. It comments on the influence techno-advancements have had upon mediating relationships in the context of the irreversible alterations taking place within the environment. It draws on the emerging capability within the robotics community to create life-like humanoid robots.

Velonaki's fascination with fictitious characters and impossible relationships led her to create another quasi-impossible analogy between an obviously fictional character, like a snowman, and a woman. Here, the snowman—as an honest representation of a fictional being—is more real than the woman. The 'woman', although she appears to be human, is an android robot, Repliee Q2, created by Professor Hiroshi Ishiguro of Osaka University. Velonaki and Rye visited Ishiguro's laboratory, where they observed and studied the behaviour of Repliee, a robot very different from those designed by our team. Velonaki and Rye, together with cinematographer Jackie Farkas, filmed Repliee using chroma key background on location. A special dress for Repliee was created by fashion designer Sin Won especially for the project. The footage of the snowman and the snowscape, where the woman was later overlaid, were both shot in the Snowy Mountains in Australia. In the installation space, the two characters—the woman and the snowman—are projected on two separate walls; one wall projection is dedicated to the snowman, and the adjoining wall to the woman. In the middle of the space, an autokinetic responsive sculpture stands between the two characters.

The autokinetic responsive object containing a small rotating screen was inspired by early spatial-kinematic mechanisms. The spherical four-bar linkage displayed selected frames in a small screen embedded in a toroidal housing, emphasising the three-dimensionality of the spatial mechanism. Displayed on the small oscillating screen are images of Swiss alpine countryside, with the same house alternately appearing snow-covered and verdant. These images are interposed with a panning shot of 'frozen' characters at a stylised formal dinner party.

Most spherical mechanisms are complicated, and hidden away in places such as the front-wheel-drives of cars, but this mechanism is minimal and elegant, containing two chrome-plated concentric semi-spherical shells created to serve as a basis for the autokinetic sculpture. The mechanism is computer-controlled to move the semi-spherical shells so that they appear to roll and turn inside each other, shifting directions backwards and forwards, stopping and resuming. The kinetic mechanism moves in various directions and speeds, acting as a 'conductor' of the installation's soundscape by selecting and mixing in real time pre-recorded compositions. Velonaki collaborated with composer Nikolas Doukas who created original compositions for the work. In juxtaposition to the mechanical appearance of the kinetic objects and minimal aesthetic of the video imagery projected on the walls, Doukas' classical compositional approach is magnificently rich and multi-faceted, employing a variety of acoustic instruments including mandolin, violin, cello, guitar and piano, together with the song of swallows.

CURRENT DIRECTIONS

The success of our collaboration led to the establishment of the Centre for Social Robotics, co-founded and co-directed by Velonaki and Rye within ACFR in 2006. With Velonaki's appointment to the National Institute for Experimental Arts (NIEA/COFA) at the University of New South Wales the collaboration has expanded to encompass a partnership between Velonaki's newly established Creative Robotics Lab at NIEA and the Centre for Social Robotics at ACFR.

Our research will continue to focus on human–robot interaction with the aim of enhancing the experience and effectiveness of interactions between people and technological others. A multidisciplinary approach is essential to achieve breakthroughs that advance the efficient, fluent use of technological systems by people. Our approach brings together and capitalises on the expertise of our established multidisciplinary team to invent new ways of working—ways that combine expertise in the creative arts, science and engineering. By conceptualising and realising projects that incorporate novel experimental interfaces, artists create unique environments that act as triggers for new behaviours to manifest, behaviours that can subsequently lead to new understandings of how people can interact with technological devices. Rapid scientific and technological progress is currently being made in robotics, and robots are beginning to move out of factories and laboratories into everyday life. The maturing of robotics as a field and the emerging awareness of the importance of social, ethological and aesthetic factors to mainstream robotics science constitutes an unprecedented opportunity to significantly advance cross-disciplinary understanding in both the creative arts and robotics.

ACKNOWLEDGEMENTS

We would like to acknowledge the substantial contributions made by our collaborators, both academic and technical, to the projects described here. Although you are too many to mention individually, you know who you are and we thank you all!

Portions of this essay were previously published.[4]

ENDNOTES

1   Lena Mamykina, Linda Candy and Ernest A. Edmonds, 'Collaborative Creativity', *Communications of the ACM* 45 (2002): 96–99.

2   Anna Andreevna Akhmatova, *The Complete Poems of Anna Akhmatova*, trans. Judith Hemschemeyer (Brookline: Zephyr Press, 1998).

3   David Silvera Tawil, David Rye and Mari Velonaki, 'Interpretation of the Modality of Touch on an Artificial Arm Covered with an EIT-based Sensitive Skin', *International Journal of Robotics Research* 31(2012): 1627–42.

4   Mari Velonaki, Stefan Scheding, David Rye and Hugh Durrant-Whyte, 'Shared Spaces: Media Art, Computing and Robotics', *ACM Computers in Entertainment* 6 (2008): 51:1–51:12.

     Mari Velonaki, David Silvera Tawil and David Rye, 'Engagement, Trust, Intimacy: Touch Sensing for Human–Robot Interaction', *Second Nature*, 2 (2010): 102–19.

     Mari Velonaki and David Rye, 'Human–Robot Interaction in a Media Art Environment' (workshop presented at the 5th ACM/IEEE International Conference on Human–Robot Interaction, 2–5 March 2010, Osaka, Japan).

3

INVENTION AT THE
SOCIAL
SCALE
[SOCIAL CONFIGURATIONS]

MELTING INTO THE TEXTURE
OF EVERYDAY LIFE

KIT MESSHAM-MUIR

ON BUILDING A PERCEPTUAL APPARATUS
— EXPERIMENTS IN PROXIMITY

JOHN TONKIN

MELTING INTO THE TEXTURE OF EVERYDAY LIFE

KIT MESSHAM-MUIR

In only the most recent years, communications technologies have evolved rapidly and now are at the centre of social life in Western culture. Technologies now respond to us in the most intuitive ways and have become omnipresent to the extent that they are becoming invisible. This ubiquitous networked culture is in turn changing our attitudes to information and knowledge, and, importantly, our expectations of our participation within them. Knowledge is now understood less as a static entity to be transmitted and more as an active relational process, taking place within social connections. This chapter looks at the ways in which recent and rapid evolutions in technology profoundly impact attitudes and expectations at a social level and, in turn, affect the ways in which knowledge is sorted and synthesised in this emerging socio-technological paradigm. In what ways might these shifts be changing the relationship between audiences and contemporary art? This chapter begins to address this question by considering the responsive works of artist John Tonkin. Based in Sydney, Tonkin's works respond to the physical presence of their audience and, in the case of his public work *Nervous System*, acts as a catalyst for social connectivity. I will suggest here that our networked culture is actually becoming less about technology and more about synthesising knowledge at the social scale.

CONNECTIVITY

Just a little over a decade ago, the internet was optional in the same way mobile phones were at the time: we 'opted-in' to using it. In 2006–07, only 67% of Australian homes had internet access,[1] and 47% of those connections were via dial-up.[2] That is, almost half the household internet usage in Australia involved making a conscious decision to dial-in through a modem from a computer. The percentage of internet access through mobile phones was in single figures[3] and largely accessed through slow and expensive pay-per-download GPRS connections, displaying stripped-down mobile websites on postage-stamp screens. The internet was thus a discrete entity, part of the optional functionality of a computer. Adam Greenfield wrote in 2006:

> A mobile phone is something that can be switched off or left at home. A computer is something that can be shut down, unplugged, walked away from. But the technology we're discussing here—ambient, ubiquitous, capable of insinuating itself into all the apertures everyday life affords it—will form our environment in a way neither of those technologies can. There should be little doubt that its advent will profoundly shape both the world and our experience of it in the years ahead.[4]

By 2011, Greenfield's prediction was reality: 'Computing has leapt off the desktop and insinuated itself in everyday life.'[5]

In 2011, 79% of Australian households had internet access,[6] with 19.5 million users (88.8% of the population).[7] Likewise in the United States: 78% of American accessed the internet in 2011; in fact, 94% of those between the ages 18–29, and 87% between the ages 30–49.[8] But the real revolution has been in how we now access the internet. Dial-up had become dinosaur technology by 2011, and 44% of all internet connections in Australia were from mobile and wireless devices. The internet is becoming less a 'computer' technology. Ownership of desktop computers is steadily declining in America (68% in 2006; 55% in 2011), while smartphone ownership continues to rise (46% in 2012).[9] As Don DeLillo's novel *Cosmopolis* prophetically suggested in 2003, 'Computers will die. They're dying in their present form. They're just about dead as distinct units. A box, a screen, a keyboard. They're melting into the texture of everyday life. This is true or not? ... Even the word computer sounds backward and dumb.'[10] The computer mutated into the smartphone and the tablet; its hard disk became the cloud and its mouse became the touchscreen. 4G mobile internet penetrates city streets, and a café without wi-fi might as well not have coffee. The internet is no longer something we log-in to; it has melted into the world.

Until recently, the 'information architectures'[11] of the internet persisted in a publishing and broadcasting model, in the 'Web 1.0' mode as we now conceive of it. In exactly the same timeframe as mobile internet has permeated social spaces, we have seen the synergetic development of the social web, Web 2.0, the internet of social networks, of Facebook, Twitter, YouTube, Tumblr, or blogging, posting and responding. According to Kathryn Zickuhr and Aaron Smith, once a person has a mobile device, their internet activity tends to increase, not just via mobile connectivity but overall.[12] In America, 67% of online adults in America use social networking;[13] in Australia, it is 58.7%[14] In fact, in 2013 49.9% of the entire Australian population has a Facebook account (11.5 million); there are 11 million YouTube accounts, and around 3 million each of Blogspot and WordPress blogging accounts, 2.7 million LinkedIn accounts, 2.6 million on Tumblr and 2.1 million on Twitter.[15] This confluence of mobile internet technologies and Web 2.0 has turned the internet into a vast array of social media.

EXPECTATIONS AND ATTITUDES

It would be a mistake to see this development of social media as simply a new communications technology. Bill Cope and Mary Kalantzis identify the changes brought about by the 'social web' as actual 'changing knowledge ecologies.'[16] They say, 'These are ideal conditions for the development of ever more finely grained areas of knowledge, cultural perspectives and localised applications of knowledge. So significant is this change, that knowledge itself may change.'[17] Accordingly, particular sets of expectations and attitudes have emerged, which have been identified in recent research.

One expectation is that information comes easy: 76% of the teachers surveyed by the Pew Research Center in the United States 'strongly agree' that internet searching has conditioned students to expect to find information quickly and easily.[18] Information is certainly abundant: at the time of writing, a Google search on the phrase 'abundance of information' returns 'About 59,600,000 results (0.27 seconds)'. Almost 60 million potential sources, retrieved in less than a third of a second: if I clicked on every result for only 12 seconds, it would still take me from now until I retire (23 years from now) to get through them, as long as I didn't leave my computer to sleep, eat and live. Of course, Google's algorithms sort this abundance of information on 'abundance of information' for me (interestingly, Wikipedia's page on 'information overload' comes up number one). In the face of this information overload, an important 'meta-skill', as George Siemens calls it, is 'the rapid evaluation of knowledge.'[19] Skim reading, once a specialist skill of overworked academics, is now a common practice, as is the capacity to quickly digest information and evaluate its potential worth.

The important shift here is from acquiring knowledge to managing it. Thus, the 'just-in-case' model of knowledge, which has hitherto been an axiom of all education, is beginning to seem less appropriate. For example, I went through the Welsh secondary education system in the 1980s. One of the routine tortures that was a supposedly vital pillar of learning was the multiplication table, or the 'times tables' as we called it. Even as rote learning was becoming seen at the time as questionable, even meaningless, learning the multiplication table was generally accepted as one exception in which rote learning was still valuable. In Mr Jones' mathematics class we recited that 'one-two-are-two, two-twos-are-four, three-twos-are-six' and so on, in a kind of perverse abridged iambic pentameter. As an inordinately cynical and admittedly lazy twelve-year-old, I could not see the point: I had a calculator if I needed to know what seven-sixes were, so the effort and time learning the 'times tables' mantra seemed like a waste. But, teachers would berate, what if I didn't have a calculator handy? I should know this *just in case*. As it happens, in the 30 years since, I've never once been in a situation where I needed to work-out a multiplication and either didn't have a calculator or wasn't with someone who had learned the multiplication table. In those situations, the information was retrieved 'just in time', not learned 'just in case'.

When information on just about any topic is available within 0.2 seconds, we are surely shifting from a 'just in case' model of knowledge to 'just in time'. Mobile devices enable this— iPhones, Androids, iPads and other smartphones and tablets. 62% of the entire adult population in the US are likely to have used the mobile internet for 'just in time' information in the last 30 days. That is, '86% of smartphone owners used their phone in the past month to make real-time queries to help them meet friends, solve problems, or settle arguments.' In fact, 27% used their smartphone to get information to settle an argument.[20] Of course, just-in-time information cannot substitute for deeper and slowly acquired knowledge, such as learning to read music or write computer code, or the manual skills of flying an airplane or playing a guitar. As John D. Cook puts it, 'The difference between just-in-case and just-in-time is like the difference between training and trying. You can't run a marathon by trying hard. The first person who tried that died. You have to train for it. You can't just say that you'll run 26 miles when you need to and do nothing until then.'[21]

Concurrent with the rise of this just-in-time knowledge-management approach to information is a shift in attitudes towards knowledge and the nature of our participation within it. Social media is turning knowledge into a social process. Knowledge is understood as less a static thing, possessed by the knowing and given to the unknowing; rather, knowledge is a creative and relational process in which information, abundant and constantly updated and shifting, is synthesised through sharing between multiple participants. In 2009, Jianwei Zhang said that Web 2.0 technologies have made it possible for 'participatory social networking and knowledge sharing, with promise for supporting knowledge creation.'[22] Thus, social networks create knowledge, not simply transmit it. The old transmission model of mediated knowledge, which has existed since the invention of the Gutenberg printing press in the mid-fifteenth century, in which authors create and readers receive, is currently giving way to an emerging participatory model, in which the roles of creator and consumer are interchangeable and audiences and authors are the same people. As Susan Cairns says, 'The core focus of Web 2.0 is participation, rather than publication.'[23] Participation takes many forms. It can mean simply posting a textual status update or tweet. And it can mean posting substantial original content, like blogs, photographs and videos: 46% of adult internet users in the United States post original photos or videos online that they have created themselves.[24]

In the grey area between creators and consumers, many social media users are also 'curators': 41% of American internet users repost photos or videos found online.[25] We're not just sharing pictures of cats, we're selecting and sharing information and knowledge and, importantly, editorialising on it. The web versions of major newspapers, such as *The Times, The New York Times* and *The Sydney Morning Herald*, carry 'share' buttons that allow us to repost to social media sites with our own comments. Fifteen per cent of Americans now get most news from personal connections on social media sites, and this 'rises to nearly a quarter among 18-to-25-year-olds.'[26] Content aggregators, such as *Google News* and *The Huffington Post*, glean news stories from multiple sources and, in re-presenting them together, fragment the authority of the traditional mainstream newspapers, which are in rapid decline.[27]

For the younger generations that dominate social media,[28] knowledge sharing is perhaps less about simple redistribution and more about a co-creative, crowd-sourcing, creative commons approach to knowledge. Cairns notes a 'swing away from dissemination and toward mutualization'.[29] This swing is perhaps registered more acutely in the field of education.[30] According to Susan D. Blum, one impact is a significant rise in plagiarism which, she argues, is not simply because the internet facilitates the easy cut-and-paste of text; rather, 'It has *changed* how [students] think of texts.'[31] Similarly, Trip Gabriel says, 'The Internet may also be redefining how students—who came of age with music file-sharing, Wikipedia and Web-linking—understand the concept of authorship and the singularity of any text or image.'[32] Maybe the emerging generations will eventually reject present understandings of intellectual property. In the earliest period of the printing press, known as the 'incunabula' (1452–1501), conventions of publishing were still in formation—tables of contents, page numbering, chapter headings and *copyright* simply did not exist.[33] It took 250 years for ideas of authorship and intellectual property to consolidate into copyright laws for the world of the printing press.[34] In 2013, we are moving rapidly through a 'digital incunabula',[35] and we cannot know at this point where these ideas will settle. Nevertheless, what we are registering is changing ideas of knowledge, moving away from knowledge as a *thing* that is held and owned to a connective relational *process*.

CONNECTIVISM

George Siemens was one of the earliest education theorists to recognise the potential ways in which the social capacities of internet technologies could radically change approaches to knowledge, creating the term 'connectivism'.[36] Bruce Neubauer, Richard Hug, Keith Hamon and Shelley Stewart say, 'Connectivism is an emerging explanation or theory of learning regarding the significance of networks (nodes and connections) as related both to individual learning and the collective generation of knowledge.'[37] Terry Anderson says, 'connectivist theorists stress the value of peer-to-peer interaction in investigating and developing multiple perspectives.'[38] In forming the theory, Siemens idenitifies three epistemological traditions in learning:

> Objectivism (similar to behaviorism) states that reality is external and is objective, and knowledge is gained through experiences. Pragmatism (similar to cognitivism) states that reality is interpreted, and knowledge is negotiated through experience and thinking. Interpretivism (similar to constructivism) states that reality is internal, and knowledge is constructed.[39]

Siemens argues that common to these models of thinking about knowledge is the tenet that learning occurs inside a person.[40] Yet, he argues, knowledge occurs between people; it is 'stored' and actively constituted within social connections.[41] And, in a networked world, 'The capacity to form connections between sources of information, and thereby create useful information patterns, is required to learn in our knowledge economy.'[42]

Connectivism, Siemens argues, is the theory that knowledge is formed when information flows between the nodes of social and technological networks, which are 'nebulous environments of shifting core elements'. Thus, knowledge is continually changing: 'While there is a right answer now, it may be wrong tomorrow due to alterations in the information climate affecting the decision.' The ability to sort and evaluate information is therefore a core skill.[43] Siemens' approach has been criticised, largely for his over-claiming that it represents a whole new learning theory.[44] Written in 2005, there is no mention of social media in Siemens' 'Connectivism: A Learning Theory for the Digital Age', but other theorists have since considered more specifically how social media can 'transform scholarship into social scholarship'.[45] In this new paradigm of knowledge, then, education is no longer defined as the single-direction transmission and reception of knowledge; instead, it becomes about creating the conditions, spaces and opportunities for knowledge to be created, to collide and to synthesise. The important thing is that the key to connectivism is not technology, but people and the connections between them, which are hyper-enabled by technology.

Technologies such as The Cube at Queensland University of Technology (QUT) assume a technologically enabled socially networked audience. The Cube, activated in 2013, is essentially three large panoramic screens wrapped around the four sides of an L-shaped structure at the QUT's Science and Engineering Centre in Brisbane. It is a publicly accessible space, which is also used for teaching and research. The screens are responsive to multiple touch, which allows many users to point, move, tap to open, pinch and zoom, and other movements which seem natural for a contemporary audience. The Cube has a number of different applications, such as a large interactive animated underwater Virtual Reef, the Physics Playroom in which certain properties such as gravity can be varied and the Flood Wall, based around the 2011 flood that inundated large areas of Brisbane's waterfront. It combines the earth science surrounding the flood with its social dimension as an experience shared by the people of Brisbane. The public can upload their photographs, videos and written pieces to the wall, place them on a timeline and a map, and see similar posts left by others. The scale of the wall opens up the virtual space into the physical and social space of The Cube—people can share stories, point to their homes, show others their photographs—so the social networking takes place on both the virtual and real sides of the screen.

NERVOUS SYSTEM

In 2011, Sydney-based artist John Tonkin completed a responsive light and sound installation called *Nervous System* in a public park for the Southport Broadwater Parklands Redevelopment in the Gold Coast, Queensland. The work consists of 20 steel poles lit with a vertical strip of LEDs that emit brightly coloured moving shapes and electronic blips. Both the sound and lights of each pole respond to light, sound, temperature and motion in their immediate environment, as well as the movement and proximity of people present. Each pole responds to its immediate proximity and is also connected to the other poles. Each function as a node in the networked whole, hence Tonkin's title *Nervous System*. Like an organic nervous system, the work senses its environment and responds. And when the social dimension is added, the work becomes reflexive.

Nervous System is removed from the immediate contextual frame of art and all that implies. Sure, aficionados of contemporary art familiar with Tonkin's work are likely to see *Nervous System* in the context of Tonkin's practice, his stated concerns with phenomenology, embodiment and thinking; but what does the work mean to a jogger, a dog walker or a group of teenagers on BMXs who don't bring this frame to the work? Watch people respond to the work, and vice versa, in the evening twilight: the ad-hoc investigations and experimentation that *Nervous System* provokes—the way the work 'plays' its audience and, in turn, their responses to each other—activate this public space as a connected social space. When its audience reaches a critical mass, *Nervous System* becomes analogous to a digital social network.

Like *Nervous System*, much of Tonkin's work responds directly to its audience. Unlike *Nervous System*, much of his work *looks* more like video art: flat screen monitors mounted on walls in a darkened gallery space. The video images on Tonkin's screens are nearly abstracted incidental moments that move just enough not to be considered as still images. Their soundtracks barely rumble and whisper. There is perhaps just enough of a visual hook to draw you closer, and it is at that moment that we realise these videos are responding to our physical movements. From the screen of Tonkin's *Selective Attention*, 2011, we hear rain pouring. The image is a green and white blur, moving minutely, until we approach it. As we close in on the screen, the narrow depth of field of the image pulls in and the focus sharpens to reveal heavy rain seen through a window. When we approach another work, *meta-cognition*, our movements fast-forward images that rush through a forest, a blurred forward view from a moving car, and the same images pictured on *other* screens. We see another figure watching these images, just like us, in a gallery space and sometimes in a surrealistic curved room. Moving back and forth in front of the work, we scrub at high speed through this montage.

Tonkin's interactive videos actually resist the designation of either 'interactive' or 'video':

> I've been really interested in exploring somewhere between those two things, something that's more like a video work but is still interactive. I coined the phrase 'responsive video'.[46]

True enough, his works do almost nothing without the engagement of their audience, yet 'interactive' no longer seems like the right term—not only for Tonkin's work, but also for the kind of engagement of today's audience. Although interactive multimedia is still relatively new, the term itself already carries certain baggage. It summons to mind the information retrieval of a touch-screen directory in a shopping mall or airport, or the didacticism of educational museum technology, such as the interactive timeline of the life of Winston Churchill at the Churchill Museum, London: projected onto a large tabletop, museum visitors touch and manipulate the screen to open years, months and dates in Churchill's life, to open files and expand images and documents. 'Interactive' implies an invitation to a particular audience to interact for a particular purpose. Interactive interfaces have often been passive and manual: a two-dimensional touch screen awaiting the hand of the user. The narrative route through many interactives can be plotted on a flow chart as a series of questions, providing one set of options that lead to another set, and so on.

Tonkin's responsive videos, on the other hand, do something different. The works set traps that lay dormant until their audience, often unknowingly, trigger a response by their physical proximity to the works. When we become aware of the connection between our own bodily movements and the movement on screen, we become consciously aware of the feedback loop that is created by our movements and the effects of the video. Even if we are familiar with Tonkin's other responsive works, we approach a new work not knowing how the work will respond to our presence. For some works, our varying proximity will cause the image to blur or sharpen; others will cut between different 'scenes'; others will scrub back and forth at different speeds. Each works responds differently, and with certain works each encounter with the same work will create a different response. There is no predictable flow chart of responses. Although it becomes quickly apparent that the works are responding to the movement of our bodies, it is not obvious how. Tonkin's audiences in the gallery space often attempt to provoke different responses by waving arms, lifting legs, walking towards the works and backing away. At this point, audiences 'play' Tonkin's work with their trial-and-error movements in space: if I move closer to the screen, the bottle on the escalator slows down. Am I causing that? What's happens when I step backward or move my arms? The audience's embodied performance in proximity to the work is necessary for the work to perform in response. Tonkin says that his audience plays his works 'like a musical instrument'.[47] Perhaps the musical instrument closest to Tonkin's work is the theremin, the kinetic electronic instrument played only by the gestures of a musician. When a musician plays a theremin, the instrument demands that they perform certain movements to create certain sounds and, in effect, the instrument 'plays' the musician. Tonkin's responsive videos likewise 'play' their audience. The roles of both the work and its audience shift into a more active performative mode.

In this way, Tonkin's responsive works reconceive interactive multimedia from its usual asymmetrical flow of information, from media to receiver, to a more collaborative synthesis. As Tonkin says, 'I've always been interested in this idea of blurring the boundary of who's making the art.'[48] The process becomes less about information retrieval and more a creative and improvisational process. Or, to put it another way, unlike the usual approach of interactive multimedia, in Tonkin's work information is not gained by the audience, but is instead generated in a process of co-creation.

Tonkin's *Nervous System* is different in many respects from The Cube at QUT. It is much more abstract and much less didactic. However, similar to The Cube, *Nervous System* generates a technologically activated social space that takes place between those engaging with the technology. Visitors to The Cube and the public that engage with *Nervous System* are practically the same people—the audience, in the broader sense, is certainly exactly the same. It's the same audience settling arguments using their iPhones, sharing news stories on Facebook, sorting information, curating their lives, posting videos and photographs. Tonkin's responsive works intuitively address certain inclinations in the audience to engage technology in a participatory and social way. Like social media, the work is *activated* within social connections as much as it activates. What happens in *Nervous System*, and in so many other emerging communication technologies, happens person-to-person as much as it does between people and technology. All of Tonkin's responsive works, to some extent, push us back out of technology and into the social space. That may be where our future with technology lies—not within virtual reality, but with the virtual in reality.

ENDNOTES

1   'Use of Information Technology 1301.0, Year Book Australia, 2012', Australian Bureau of Statistics. Accessed 30 April 2013, www.abs.gov.au/ausstats/abs@.nsf/Lookup/by%20 Subject/1301.0~2012~Main%20Features~Use%20of%20information%20technology~174.

2   Summary, 8153.0, Internet Activity, Australia, June 2006', Australian Bureau of Statistics. Accessed 30 April 2013, www.abs.gov.au/AUSSTATS/abs@.nsf/allprimarymainfeatures/ D7D9E5103E439E3FCA2572830017E4C5?opendocument.

3   'Use of Information'.

4   Adam Greenfield, *Everyware: The Dawning Age of Ubiquitous Computing* (Berkeley: New Riders, 2006), 6.

5   Greenfield, *Everyware*, 9.

6   'Use of Information'.

7   'Oceana and South Pacific', Internet World Stats, 2011. Accessed 6 May 2013, internetworldstats.com/pacific.htm#au .

8   Kathryn Zickuhr and Aaron Smith, 'Digital Differences: While Increased Internet Adoption and the Rise of Mobile Connectivity Have Reduced Many Gaps in Technology Access over the Past Decade, for Some Groups Digital Disparities Still Remain' (Pew Research Center's Internet & American Life Project: Washington DC, 2012), 5. Accessed 30 April 2013, pewinternet.org/ Reports/2012/Digital-differences.aspx.

9   Zickuhr and Smith, 'Digital Differences', 14.

10 Don DeLillo, *Cosmopolis* (New York: Scribner, 2003), 46.

11 Bill Cope and Mary Kalantzis, 'The Role of the Internet in Changing Knowledge Ecologies', *Canadian Journal of Media Studies,* 7 (2010): 1–23.

12 Zickuhr and Smith, 'Digital Differences', 14.

13 Maeve Duggan and Joanna Brenner, *The Demographics of Social Media Users—2012.* (Pew Research Center's Internet & American Life Project: Washington DC, 2013). Accessed 6 May 2013, pewinternet.org/Reports/2013/Social-media-users.aspx.

14 David Cowling. 'Social Media Statistics Australia—April 2013', *Social Media News.* Accessed 6 May 2013, www.socialmedianews.com.au/social-media-statistics-australia-april-2013/; 'Oceana and South Pacific.'

15 David Cowling, 'Social Media Statistics Australia—April 2013.'

16 Cope and Kalantzis, 'The Role of the Internet'.

17 Cope and Kalantzis, 'The Role of the Internet', 5.

18 Kristen Purcell, Lee Rainie, Alan Heaps, Judy Buchanan, Linda Friedrich, Amanda Jacklin, Clara Chen, and Kathryn Zickuhr. 'How Teens Do Research in the Digital World: A Survey of Advanced Placement and National Writing Project Teachers Finds that Teens' Research Habits Are Changing in the Digital Age' (Washington DC: Pew Research Center's Internet & American Life Project, 2012), 3. Accessed 30 April 2013, pewinternet.org/Reports/2012/Student-Research.

19 George Siemens, 'Connectivism: A Learning Theory for the Digital Age', *International Journal of Instructional Technology and Distance Learning,* 2(1), (2005): 5.

20   Lee Rainie, Joanna Brenner and Kristen Purcell (2012). *Photos and Videos as Social Currency Online* (Washington DC: Pew Research Center's Internet & American Life Project, 2012), 4. Accessed 1 May 2013, pewinternet.org/Reports/2012/Online-Pictures.aspx .

21   John Cook, March 16, 2010, 'Emacs', *The Endeavour*, Accessed 30 April, 2013. www.johndcook.com/blog/page/79.

22   Jianwei Zhang, 'Toward a Creative Social Web for Learners and Teachers'. *Educational Researcher*, 38 (2009), 276.

23   Susan Cairns, 'Mutualizing Museum Knowledge: Folksonomies and the Changing Shape of Expertise', *Curator: The Museum Journal*, 56 (2013):108.

24   Rainie, Brenner and Purcell, 'Photos and Videos'.

25   Rainie, Brenner and Purcell, 'Photos and Videos', 2.

26   Pew Research Center, *Internet & American Life Project* (Washington DC, 2013). pewinternet.org/Reports/2013.

27   Matthew Knott, 'Circulation Figures: The Paper Is Dead, Long Live the "Masthead" ', *Crikey*. 15 February 2013. Accessed 6 May 2013, www.crikey.com.au/2013/02/15/circulation-figures-the-paper-is-dead-long-live-the-masthead/; The Pew Research Center, *The State of the News Media 2012: An Annual Report on American Journalism*. (The Pew Research Centre's Project for Excellence in Journalism: 2012). Accessed 6 May 2013, stateofthemedia.org.

28   Duggan and Brenner, *The Demographics of Social Media*, 3; Lenhart, 'Teens & Online Video: Shooting, Sharing, Streaming and Chatting—Social Media Using Teens Are the Most Enthusiastic Users of Many Online Video Capabilities', (Washington DC: Pew Research Center's Internet & American Life Project, 2012). Accessed 6 May 2013, pewinternet.org/Reports/2012/Teens-and-online-video.aspx.

29   Cairns, 'Mutualizing Museum Knowledge', 107.

30   Siemens, 'Connectivism'; Cope and Kalantzis, 'The Role of the Internet'; Zhang, 'Toward a Creative Social Web'; Christine Greenhow, Beth Robelia and Joan Hughes, 'Research on Learning and Teaching with Web 2.0: Bridging Conversations', Education Researcher, 38(2009); Bruce J. Neubauer, Richard W. Hug, Keith H. Hamon and Shelley K. Stewart, 'Using Personal Learning Networks to Leverage Communities of Practice in Public Affairs Education', *Journal of Public Affairs Education*, 1(2011): 9–25.

31   Susan Blum, 2009, original emphasis. *My Word! Plagiarism and College Culture* (New York: Cornell University Press, 2006), 5.

32   Trip Gabriel, 'Plagiarism Lines Blur for Students in Digital Age', *The New York Times*. 1 August 2010. Accessed 6 May 2013 www.nytimes.com/2010/08/02/education/02cheat.html?_r=1.

33   Cope and Kalantzis, 'The Role of the Internet', 4.

34   Jean-Claude Guédon, *In Oldenburg's Long Shadow: Librarians, Research Scientists, Publishers and the Control of Scientific Publishing*. (Washington DC: Association of Research Libraries, 2001), 11.

35   Cope and Kalantzis, 'The Role of the Internet', 4.

36   Siemens, 'Connectivism'.

37   Neubauer, Hug, Hamon & Stewart, 9.

38   Terry Anderson, *The Theory and Practice of Online Learning* (Edmonton: AU Press, 2008), 57.

39   Siemens, 'Connectivism', 4.

40   Siemens, 'Connectivism', 5.

41   Siemens, 'Connectivism', 3.

42   Siemens, 'Connectivism', 6.

43   Siemens, 'Connectivism', 7.

44   Anderson, *The Theory and Practice of Online Learning*, 18.

45   Greenhow, Robelia and Hughes, 'Research on Learning and Teaching', 1.

46   John Tonkin, artist interview, 21 April 2013.

47   Tonkin, 2013.

48   Tonkin, 2013.

ON BUILDING A
PERCEPTUAL APPARATUS —
EXPERIMENTS IN PROXIMITY

JOHN TONKIN

The relationship between the mind and the body has been contested for millennia. Descartes regarded the body and senses with suspicion, and this separation between mind and body still seems to underpin many strands of contemporary thought. Historically, dominant viewpoints in the philosophy of mind and cognitive science have considered the body as peripheral to the understanding of the nature of mind and cognition. The last century has seen a growing acceptance of the significance of the body, initially in fields such as pragmatist philosophy, phenomenology, cybernetics, and ecological psychology as well as those more recent fields that more obviously fall under the rubric of embodied cognition. Embodied cognition is made up of a broad collection of approaches that acknowledge that minds are embodied within a physical body and embedded within a physical environment.

Over the last three years I have been building a number of responsive video artworks that investigate embodied cognition. These works sit under a collective title *Experiments in Proximity,* which relates to the underlying interactivity that is controlling these works but also to the idea of proximity as a concept for thinking about how we perceive and act in the world. They are part of a broader practice-led research project that could be thought of as a meta meta-cognition, for which I have given the working title 'thinking about how we think about thinking'.

> I am much more fascinated in how we think about how we think, than I am in the actual process of thought itself. These attitudes and beliefs about thinking underlie how we inhabit both ourselves and the world.[1]

In this chapter I will examine a number of ideas that have emerged from embodied cognition that might offer insights into the interactive artworks that I have been making, positioning them within this networks of ideas. This discussion circulates around a set of artworks commencing with *Closer: Eleven Experiments in Proximity* (2010), the first in the series of works that I will discuss.

Closer sets up the initial configuration of the works that the later works build upon. The basic unit is a large flat screen attached to the wall along with a sensor that is able to determine the physical proximity of a viewer from the screen. The system responds to this distance by dynamically updating the sound and video content that is shown on the screen. Each of the experiments consists of a different set of couplings between movement and perception. The imagery is grounded in everyday lived experience and hints at the poetic breadth of simple moments of engagement with the world; walking along a street, moving closer to a sleeping lover, a kettle boils, a discarded water bottle spins at the bottom of an escalator, slowing to a standstill as you approach. The viewer is located within a kind of first person temporal play that mirrors our daily navigation of distance—both physical and emotional:

> … the durational dynamics of each scene are intensified through Tonkin's placement of interactive sensors within the installation. Building on his ongoing interest in the interstices of time that make up our cumulative experience of duration and motion, the work choreographs the viewer's body against the changing speed of the scenes as they play. Running or walking, advancing or retreating in proximity to the work serves to change the tempo and narrative progression of each video, from fast to slow, forwards or backwards, respectively. By using the audience as a wayward metronome, Tonkin highlights the body's liminality in relation to time's unfolding.[2]

THE EMERGENCE OF EMBODIED COGNITION

Cybernetics was defined by Norbert Wiener in 1948 as the science of communication and control in the animal and the machine.[3] It used a language of circuits, feedback, and information flow to describe how systems, both organisms and machines, functioned. It grew out of control theory and systems theory to influence a very broad range of fields including information theory, biology, ecology, sociology, anthropology, psychology and philosophy. Cybernetics was instrumental in the development of the fields of artificial intelligence and cognitive science, and tends to feature prominently in histories of computing. More recently, a number of alternatives to these canonical histories of computers have been published. Paul Edwards presents a history of computers as a technological development central to the Cold War and as a key metaphor in psychological theory.[4] He argues that what people have experienced through computers as reflections of their own minds and as icons of rationality is as significant as anything that computers have actually accomplished. Jean-Pierre Dupuy suggests that the development of cybernetics represented not so much the anthropomorphisation of the machine as the mechanisation of the human.[5] This is perhaps most obviously made manifest in the computational theory of mind (CTM) which is commonly embraced in philosophy of mind, cognitive science and cognitive psychology. CTM regards the mind as an information processing system and thought as a form of computation. Thought is modelled as progressing step-wise from perception to cognitive processing to action. In this model, cognition is sandwiched between two layers—sensory input and motor output—which are not in themselves considered as cognitive. Cognition is seen as existing solely in the brain.

Katherine Hayles' history of cybernetics focuses on the fate of embodiment and the cybernetic machine.[6] She traces how 'information lost its body, that is, how it came to be conceptualised as an entity separate from the material forms in which it is thought to be embedded'.[7] Hayles goes on to look at how cybernetics was further developed into what came to be known as 'new cybernetics' or second-order cybernetics, by practitioners who were trying to reconcile cybernetics with complex natural open systems. This included researchers such as Gregory Bateson and biologists Humberto Maturana and Francisco Varela. Many of the ideas that emerged from this 'new cybernetics' have firmly reconnected cognition back to the physical body and have gone on to underpin much subsequent thinking about embodiment and embeddedness.

In 1970 Maturana wrote that 'Living systems are cognitive systems, and living as a process is a process of cognition. This statement is valid for all organisms, with or without a nervous system'.[8] Cognition, in this approach, is extended beyond a question of mind and/or symbolic processing to being a process that also operates at other levels of the organism. Maturana and Varela went on to develop a theory of autopoiesis about the nature of reflexive feedback control in living systems.[9] It was originally used to describe the interplay between the structure and function of the sub-components of biological cells. They described the ongoing operational interrelationships between an organism and its environment as structural coupling and went on to expand these ideas of structural coupling and circular causality to encompass perception, cognition and consciousness.[10] This set a platform for more radical approaches to the mind such as extended mind[11] and enaction.[12] I am going to focus on the ideas of structural coupling enaction and how they might be applied to thinking about the works in *Experiments in Proximity*.

STRUCTURAL COUPLING AND ENACTING THE WORLD

Enaction is the idea that organisms create their own experience through their actions. Organisms are not passive receivers of input from the environment, but are actors in the environment such that what they experience is shaped by how they act.[13]

In 1966 Jerome Bruner proposed three modes of representation in child development: enactive representation (action-based), iconic representation (image-based), and symbolic representation (language-based).[14] In 1991 Varela, Thompson and Rosch, building on the work of Maturana and Varela, adopted the term enaction to describe cognition as being grounded in the ongoing dynamic interactions between a living organism and its environment.[15] Rather than CTM's idea of information progressing linearly from sensory input to cognition to motor output, they proposed that 'sensory and motor processes, perception and action, are fundamentally inseparable in lived cognition. Indeed, the two are not merely contingently linked in individuals; they have also evolved together'.[16] Instead of the informatics problem of recovering pre-existing properties of the world,[17] they suggest that we bring forth or enact the world through our active bodily engagement with it.

An often used example of structural coupling between sensory and motor processes is in perceiving the softness of a sponge.[18] The softness of a sponge is found in its dynamic response to the active probing and squeezing of a curious perceiver. Softness is not a property of the sponge that exists in isolation to this interaction.

Our first-person visual experience of the world is of a rich, detailed and colourful world—a direct translation of the image projected onto the retina. Modern theories of visual perception have had to account for evidence that has emerged from empirical studies that the visual apparatus is far from camera-like. For example, spatial resolution is not constant across the retina. The fovea—which corresponds to the sharp central region of the visual field—is small, comprising less than 1% of the retinal surface, yet it feeds more than half of the visual cortex. Colour resolution also drops off rapidly when moving out from the fovea to the peripheral retina.

> Despite these non-homogeneities, the perception of spatial detail and color does not subjectively appear nonuniform to us: most people are completely unaware of how poor their acuity and their color perception are in peripheral vision.[19]

There are many other 'problems' with the visual system (for example, blurring caused by saccadic motion and the blind spot) that also need to be accounted for. Modern theories of perception that have grown out of a CTM perspective (see, for example, David Marr[20]) posit that perceptual experience in humans depends solely on the activation of internal representations. Within these theories it is thought that it is these representations that fill in the gaps caused by the limitations of the visual system.

O'Regan and Noë counter these representationalist models with their sensorimotor account of visual perception[21] that shares much in common with the enactive approach. They suggest that the world serves as its own external representation and that perception consists of the bodily exploration of the world. Within their approach, the act of seeing is physically active; we maintain the sense of a coherently present visual field by visually attending to the world in a manner that is much like how we think of touching something—we reach out and we touch it:

> ... as soon as you do cast your attention on something, you see it ... One could make the amusing analogy, referred to by Thomas (1999)[22], of the refrigerator light. It seems to be always on. You open the refrigerator: it's on. You close the refrigerator, and then open it again to check, the light's still on. It seems like it's on all the time! Similarly, the visual field seems to be continually present, because the slightest flick of the eye, or of attention, renders it visible.[23]

PERCEPTUAL APPARATUS AND ENACTIVE MACHINES

I would like to posit that the responsive video systems that I have been building for *Experiments in Proximity* might be considered as enactive machines. A viewer is inserted into an interrelationship with these perceptual apparatus that could be considered what Simon Penny describes as a 'prosthetic extension of their embodiment'.[24] To begin my elaboration of this claim, I will describe another artwork from the series that is titled *Selective Attention* (2011).

Selective Attention is perhaps the simplest of the responsive video works. From a distance the viewer sees an ambiguous blurred form. On approach, the edge of a weathered window frame comes sharply into focus; streaks of rain become discernible through the window, as does the sound of the rain. Moving even closer to the screen pushes the plane of focus (that part of the image where the image is sharp) out through the window and into the rain. The sound of the now obviously very heavy rain becomes louder than ever. By moving backwards and forwards in front of the work, the viewer is able to, quite literally, focus on different parts of the scene. There is an embodied physical tangibility to this experience that might not be so apparent in reading a description.

This particular work was created using a single piece of video footage that consists of a slow continuous focus pull whereby the plane of focus moves continuously from the camera to the window and into the far distance. The viewer's proximity to the screen is determined by a Kinect sensor (typically between a range of one and four metres). This proximity value is used to determine what frame from the footage to show, as well as the volume of the sound. The sensor is very sensitive; moving a couple of centimetres is enough to move forward or backwards through the footage. Often there are refinements that need to be made that are specific to each work. In this case a small amount of stochastic jitter is added to the proximity value. This means that even when the viewer is stationary the footage jumps around a little, making the rain appear to fall rather than being frozen in place.

Drawing from the language that I have laid out in the last few pages, a dynamic structural coupling is formed between the viewer and the interactive system. The input into the interactive system is the viewer's proximity to the screen; the interactive system responds to this and updates the audio-visual content on the screen. This, in turn, becomes the input to the viewer, who might respond by moving in relationship to the screen. The cybernetic loop is, in a sense, closed by the viewer's curiosity and desire to make sense of the enactive machine within which the viewer has become embedded. To engage with the work, the viewer needs to extract an emergent model of this sensorimotor apparatus. How is the input (proximity) mapped to the outputs (audio and video)? The answer could arise through conscious thinking or a more tacit embodied understanding of how the system is working:

> … the persuasiveness of interactivity is not in the images per se, but in the fact that bodily behavior is intertwined with the formation of representations. It is the ongoing interaction between these representations and the embodied behavior of the user which makes such images more than images. This interaction renders conventional critiques of representation inadequate and calls for the theoretical and aesthetic study of embodied interaction.[25]

It is this embodied interaction or enactive knowledge—expressed through the dynamics of the interactivity itself—that is my primary interest in building these responsive video systems. The actual audio-visual content for the works is secondary to this goal. In order to highlight the primacy of the interaction, I have tended to use understated content that draws from small moments of everyday lived experience. This footage has mostly been shot from a first-person perspective and creates the sense of a subjective, phenomenological engagement with the world.

SENSORIMOTOR CONTINGENCIES, KITTENS AND A VILLAINOUS AQUATIC MONSTER

I want to further unpack the expression that I used above: 'to extract an emergent model of this sensorimotor apparatus'. To do this I will explain O'Regan and Noë's concept of sensorimotor contingencies.[26]

O'Regan and Noë write about how we learn to see.[27] Within their approach, perception involves an exploration of the world that is mediated by learnt sensorimotor contingencies. A sensorimotor contingency describes a specific relationship or structural coupling between an action and a perception. For different modes of perception there are different sets of sensorimotor contingencies. So, for example, turning one's head to the right has a different effect on one's visual field (it shifts to the left) than it does on one's audio perception (there is a shift in the relative volumes presented to the left and right ears).[28]

On the theory of sensation presented here, to have a sensation is to exercise one's mastery of the relevant sensorimotor contingencies and in this sense to be 'attuned' to the ways in which one's movements will affect the character of input. We characterise this attunement as a form of practical knowledge.[29]

Varela et al. give an example of how learning to see requires being able to act in the world.[30] Held and Hein[31] raised pairs of kittens in the dark and exposed them to light only under controlled conditions during which time they were attached to a kind of carousel or gantry. For each pair, kitten A could move around relatively freely whilst harnessed to the gantry. Kitten B was attached to the other end of the gantry. It was in a basket so that it couldn't control its own movements; instead, its movements exactly mirrored those of kitten A. The two kittens effectively shared the same visual experience, but kitten B was entirely passive. When not in the gantry, both kittens were housed in darkness with their mother and siblings. After several weeks of spending three hours per day in the gantry, the A team kittens behaved normally, whereas the B team kittens behaved as if they were blind, bumping into objects and falling off edges. For kitten B, who is unable to actively move around while seeing, the appropriate visual sensorimotor contingencies are not formed.

O'Regan and Noë use a more dramatic metaphor to explain the process of learning sensorimotor contingencies that I think is useful. One could describe the activity of the engineers on the underwater vessel as having to extract an emergent model of their sensorimotor apparatus.

> Imagine a team of engineers operating a remote-controlled underwater vessel exploring the remains of the Titanic, and imagine a villainous aquatic monster that has interfered with the control cable by mixing up the connections to and from the underwater cameras, sonar equipment, robot arms, actuators, and sensors. What appears on the many screens, lights, and dials, no longer makes any sense, and the actuators no longer have their usual functions. What can the engineers do to save the situation? By observing the *structure of the changes* on the control panel that occur when they press various buttons and levers, the engineers should be able to deduce which buttons control which kind of motion of the vehicle, and which lights correspond to information deriving from the sensors mounted outside the vessel, which indicators correspond to sensors on the vessel's tentacles, and so on.[32]

Returning to the first work that I described, *Closer: Eleven Experiments in Proximity*, this work is made up of eleven different experiments in the coupling of proximity to sound and vision. Each experiment plays out for 30 seconds before moving onto the next. In a sense there are eleven different sets of sensorimotor contingencies to be deciphered by the viewer. Some are very literal and almost act as instructional aids for the viewer; using familiar technologies (an automatic door, an escalator) to give clues as to what to do. Some use a linear coupling between proximity and the video footage in the same way as described for *Selective Attention*. Others use more complex couplings using higher order derivatives such as the velocity or acceleration of the viewer. The sound component has a simultaneous but different coupling relationship to the proximity.

EXPERIENCES OF THE WORK

In order to highlight the primacy of the interactivity in these experiences I have tended to use understated audio-visual content that draws from small moments of everyday lived experience. To create the sense of a subjective, phenomenological engagement with the world, this footage has mostly been shot from a first-person perspective. In the more recent works, I have experimented with using a range of more complex content and narrative

Figure 3.1: *Experiments in Proximity*, 2013, Breenspace, Sydney. Photo courtesy of the Artist.

Figure 3.2: *Experiments in Proximity*, 2013, Breenspace, Sydney. Photo courtesy of the Artist.

Figure 3.3: *Experiments in Proximity*, 2013, Breenspace, Sydney. Photo courtesy of the Artist.

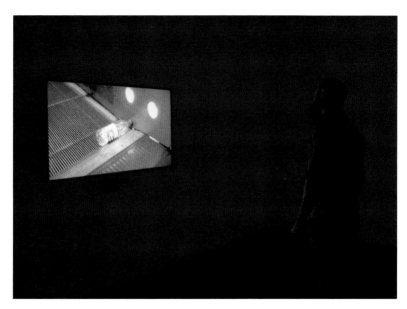

setups that engage more broadly with ideas around thinking and cognition. The work titled *A Biology of Cognition* (2011) is a speculative visualisation of Maturana's famous paper, 'Biology of Cognition'.[33] As part of Maturana's extension of the notion of cognition to encompass all of life, he presents a portrait of the phenomenological first-'person' experience of a single celled organism as it encounters a nutrient gradient. My version uses macro footage of particulate matter suspended in water to create the experience of tumbling through a microscopic world. Approaching the screen instigates contact with other organisms (people) that co-habit this world. The work *meta-cognition* (2011) explores ideas of internal representation, the homunculus and infinite regression. The 'homunculus problem' is a critique of representationalist theories of vision that involve the creation of internal representations. Who is the observer of these mental images within the brain? Some of these theories assume the presence of an internal viewer or homunculus (little man) who observes these internal representations. How is the homunculus able to see? The obvious answer is that there is another homunculus inside the head of the homunculus, which sets up an infinite regression of homunculi inside heads *ad infinitum. Meta-cognition* again uses first-person footage of small moments in the world, but in this case these moments are re-recorded and re-re-recorded (etc.) with third person views of the protagonist (me) reviewing the footage on various media (smartphone, tablet, television etc.). In this deliberately confusing work, the viewer's movement backwards and forwards navigates up and down the levels of recursive re-presentation.

The works are not presented with instructions; the encounter of the naïve viewer with the enactive machine and the self-discovery of its workings (or sensorimotor contingencies) are at the core of the experience. As a media artist, I am working from a different set of assumptions and motivations than if I were a cognitive scientist or psychologist. I haven't felt the need to systematically test the behavioural responses of the viewer. That said, I have been known to spy on viewers to ascertain how they proceed. Some viewers don't realise that the work is interactive and don't approach the work. (They must think that it is a very minimal video artwork.) Some viewers might know that the work is interactive but are not inclined to engage—they don't have the requisite curiosity that I mentioned earlier. Other viewers actively engage with the work in a playful and exploratory fashion. What I particularly enjoy witnessing is the moment when viewers first realise that a work is responding to them. I have watched a child standing in front of the work looking bored, slightly swaying. After 20 or so seconds of watching the video moving a few frames forwards and backwards over and over, he suddenly realised that the work was responding to his subtle motion and began to explore.

I have explored the use of the concepts of structural coupling, enaction, and sensorimotor contingencies to think about the specific media artworks I have been creating to facilitate

Figure 3.4: *Experiments in Proximity*, 2013, Breenspace, Sydney. Photo courtesy of the Artist.

Figure 3.5: *Experiments in Proximity*, 2013, Breenspace, Sydney. Photo courtesy of the Artist.

Figure 3.6: *Experiments in Proximity*, 2013, Breenspace, Sydney. Photo courtesy of the Artist.

self-reflection around cognition. I'd like to briefly touch upon the use of these concepts to think more broadly about the process of invention itself. The circulation of these ideas through wide-ranging fields, from philosophy to cognitive science to media arts, has two important effects. In the first instance, the circulation itself has productive and inventive effects, wherein each discipline translates and interprets and creates new meanings for these concepts. The second effect is that these ideas produce mechanisms for thinking (cognition), for thinking about thinking (psychology and cognitive science), and for thinking about thinking about thinking (as in my own artworks). For media artists such as myself, these concepts can form the basis for the building of tools, the building of understandings, and the building of experiences. In practice-led research, the relationship between the development of ideas and process is complex; the inventive process doesn't progress neatly in a linear stepwise fashion or even in a simply iterative fashion. Within this context it is perhaps more useful to think of invention in terms of a structural coupling or co-emergence of theory and praxis, a thinking that is firmly grounded in the physical process of making.

ENDNOTES

1   Tonkin, 'BREENSPACE: John Tonkin—Artist Statement.' 2013

2   Tanya Petersen, 'Somnambulant Sensing', *RealTime* 100 (2010): accessed June 30, 2013. www.realtimearts.net/article/100/10128.

3   Norbert Wiener, *Cybernetics: Or, Control and Communication in the Animal and the Machine* (New York: The Technology Press, John Wiley & Sons, 1948).

4   Paul Edwards, *The Closed World: Computers and the Politics of Discourse in Cold War America* (Cambridge: MIT Press, 1997).

5   Jean Pierre Dupuy, *The Mechanization of the Mind: On the Origins of Cognitive Science*, trans. M. B. De Bevoise (Princeton: Princeton University Press, 2000).

6   N. Katherine Hayles, *How We Became Posthuman: Virtual Bodies in Cybernetics, Literature, and Informatics* (Chicago: University of Chicago Press, 1999).

7   Hayles, *How We Became Posthuman*, 2.

8   Humberto R. Maturana and Francisco J. Varela, *Autopoiesis and Cognition: The Realization of the Living* (Boston: D. Reidel Pub. Co., 1980), 5–58.

9   Maturana and Varela, *Autopoiesis and Cognition*.

10 Humberto R. Maturana and Francisco J. Varela, *The Tree of Knowledge: The Biological Roots of Human Understanding*, Revised Edition (Boston: Shambhala, 1987).

11 Andy Clark and David Chalmers, 'The Extended Mind', *Analysis* 58 (1998): 7–19.

12 Francisco J. Varela, Evan T. Thompson, and Eleanor Rosch, *The Embodied Mind: Cognitive Science and Human Experience* (Cambridge: MIT Press, 1993).

13 John R. Stewart, Olivier Gapenne, and Ezequiel A Di Paolo. *Enaction: Towards a New Paradigm for Cognitive Science.* (Cambridge: MIT Press, 2010), 428.

14 Jerome Bruner, 'On Cognitive Growth I & II.' *In Studies in Cognitive Growth: a Collaboration at the Center of Cognitive Studies*, ed. Jerome Bruner, Rose R. Olver, et. al. (New York: Wiley, 1967), 1–67.

15 Varela, Thompson and Rosch, *The Embodied Mind*.

16 Varela, Thompson and Rosch, *The Embodied Mind*, 173.

17 Varela, Thompson and Rosch, *The Embodied Mind*.

18 Erik Myin, 'An Account of Color Without a Subject?' *Behavioral and Brain Sciences* 26 (2003): 42–43.

19 John Kevin O'Regan and Alva Noë, 'A Sensorimotor Account of Vision and Visual Consciousness', *Behavioral and Brain Sciences* 24, no. 5 (2001): 939–72 (951).

20 David Marr, *Vision: A Computational Investigation into the Human Representation and Processing of Visual Information* (New York: Henry Holt and Co., Inc., 1982).

21 O'Regan and Noë, 'A Sensorimotor Account of Vision and Visual Consciousness.'

22 Nigel Thomas, 'Are Theories of Imagery Theories of Imagination? An Active Perception Approach to Conscious Mental Content', *Cognitive Science* 23 (1999): 219.

23 O'Regan and Noë, 'A Sensorimotor Account of Vision and Visual Consciousness', 947.

24 Simon Penny, 'Trying to Be Calm: Ubiquity, Cognitivism, and Embodiment.' In *Throughout: Art and Culture Emerging with Ubiquitous Computing*, ed. Ulrik Ekman (Cambridge: MIT Press, 2012), 273.

25 Simon Penny, 'Representation, Enaction and the Ethics of Simulation', In *FirstPerson: New Media as Story, Performance, and Game*, ed. Noah Wardrip-Fruin and Pat Harrigan (Cambridge: MIT Press, 2004), 83.

26 O'Regan and Noë, 'A Sensorimotor Account of Vision and Visual Consciousness.'

27 Kevin O'Regan and Alva Noë, 'What It Is Like to See: A Sensorimotor Theory of Perceptual Experience', Synthese 129 (2001): 79–103.

28 This is a huge oversimplification on my behalf of the spatial perception of sound.

29 O'Regan and Noë, 'What It Is Like to See', 84.

30 Varela, Thompson and Rosch, *The Embodied Mind*, 174.

31 Richard Held and Alan Hein, 'Movement-produced Stimulation in the Development of Visually Guided Behavior,' *Journal of Comparative and Physiological Psychology* 56(1963): 872–876.

32 O'Regan and Noë, 'A Sensorimotor Account of Vision and Visual Consciousness', 940.

33 Maturana and Varela, *Autopoiesis and Cognition*, 5–58.

4

INVENTION AT THE SOCIAL SCALE

[CULTURAL AND SOCIO-SPATIAL CONFIGURATION]

THE ARTIST-RUN INITIATIVE: AN AGENT
THAT BLURS THE STUDIO, LABORATORY
AND EXHIBITION SPACE, CREATING
A SITE FOR INVENTIVENESS

BRAD BUCKLEY
JOHN CONOMOS

ICAN: REINVENTING THE AUTONOMY
OF THE ARTIST-RUN INITIATIVE

ALEX GAWRONSKI

THE ARTIST-RUN INITIATIVE: AN AGENT THAT BLURS THE STUDIO, LABORATORY AND EXHIBITION SPACE, CREATING A SITE FOR INVENTIVENESS

BRAD BUCKLEY
JOHN CONOMOS

This chapter will consider why much contemporary visual art, which is reflexive, innovative and questioning, and at times intentionally provocative, is being produced and exhibited in artist-run initiatives (ARIs)—sometimes referred to as alternative spaces in Canada and the US—outside the established or mainstream site of the museum. As Dutch curator and theorist Henk Slager reminds us, 'the paradigm of the public exhibition was formulated at the time of the French revolution in the 18th century', and since that time the relationship between artists, curators and museums has undergone various permutations, and has been influenced by a spectrum of factors, from commerce to morality and politics.[1]

Contemporary art is now caught between being entertainment for the general public and serving various masters, including museum trustees and directors, curators, the new breed of collector/benefactor, corporate funders and state and federal governments and their agencies, such as the Australia Council for the Arts. All these people have agency over what is made and exhibited in the museum. We would therefore speculate that it is in part the unregulated nature of ARIs, which are not saddled with the crushing ascendancy of corporate managerialism, and the fact that they most often operate geographically, economically and philosophically at the fringe of the established art world, that allows them to be a site of invention.

To understand the historical impact and the ongoing centrality of ARIs to inventiveness in contemporary art, it is important to know that unlike other forms of research or knowledge creation and transfer, whether it be in the hard sciences, the humanities or architecture, ARIs generally begin as grass roots organisations with individuals or small or large groups operating them *pro bono*.

ARIs have a long history, and evolved from precursors that included collectives of artists interested in working with experimental ideas in the 1960s, outside the mainstream gallery and museum system. Often these artists were driven by a particular set of aesthetic, political or social concerns, such as a rejection of the art object that could be traded in the marketplace (commercial galleries), feminism and the women's movement, social justice and the civil rights movement in the United States.

Some of the earliest ARIs developed in New York: the Studio Museum in Harlem and 98 Greene Street during the 1960s, followed by a rapid expansion during 1970s with Franklin Furnace, P.S. 1 Contemporary Art Center, Electronic Arts Intermix (EAI), The New Museum and Group Material to name a few.[2] In Australia, Sydney had the seminal forerunner, One Central Street Gallery, founded in 1966 by Tony McGillick, the Sculpture Centre, established around 1970, and Inhibodress gallery, established by Mike Parr, Peter Kennedy and Tim Johnson in 1971. In Adelaide the Experimental Art Foundation (EAF) was established by Donald Brook and Ian North (among others) in 1974, with the Irish activist and curator Noel Sheridan as inaugural director.

However, it is over the past two decades that ARIs have flourished beyond our capital cities, in many rural towns and remote communities, because of a variety of economic, cultural and historical reasons and the needs of those communities. More recent ARIs may be indicative of the increasing professionalism of the visual arts, due in part to the amalgamation of our art schools with universities.[3] Given the increasing emphasis on DIY (do it yourself) activity and on creativity, innovation and experimentation, recent art school graduates are searching for their own affordable 'white cube' spaces for making and exhibiting their art. As has been the case since the 1960s, artists search out the less glamorous post-industrial neighbourhoods in cities and towns. Some of these ARIs, such as Firstdraft, established in 1986, have a capacity to reinvent themselves every few years by changing the group of artists who organise the ARI. Others, such as Art Unit, which was established in 1981 by Juilee Pryor and Robert McDonald, Peloton, which closed in 2013, and SNO, 55 Sydenham Road and the recent arrival Marrickville Garage, all located in Sydney, ebb and flow, responding to the changing cultural, material and technological developments and needs of artists. The same occurs with those in rural towns and cities, such as Kandos Projects, situated in the Capertee Valley, about three and half hours west of Sydney, and Raygun Projects in Toowoomba.[4]

ARIs have also had an impact on urban planning and renewal—or, to put it another way, the reinvention of neighbourhoods. While not necessarily subscribing to US urban studies theorist Richard Florida's concept of 'the creative classes', there does seem to be a correlation between the arrival of ARIs in a post-industrial neighbourhood, or post-Fordist cities and towns, and the beginnings of urban renewal. Perhaps the most famous example would be the gentrification of SoHo (SOuth of HOuston (Street)) in New York City during the early 1970s, as artists moved into the historic cast iron buildings originally designed as commercial lofts for manufacturing, drawn by the high ceilings and large open spaces. They began to organise their own exhibition spaces to show their work. Of course gentrification eventually means that rents rise, and then artists and ARIs seek out new neighbourhoods, leaving behind the fashionable shops, restaurants and expensive apartments.

Closer to home, Kandos Projects is located in the small country town of Kandos, which was known primarily as the town that supplied the cement that built Sydney. When the cement works closed recently, the town's economy and industrial base was lost, so Kandos Projects decided to host 'Cementa_13', a biennale of contemporary art. This event brought together artists, performers and an audience that is helping to revitalise and transform the town. In this context the ARI is a site of inventiveness and a driver of urban renewal.

Art is, according to Oscar Wilde's famous description, 'useless'.[5] This notion has become one of the foundational features of art as experimentation and inventiveness. ARIs as spaces of innovation have had an impact on the local and international art world not only in relation to museums, galleries, art schools, archives and collectors; they have also allowed the manifestation of a most critical condition of art to flourish. Namely, they have permitted art to include events, processes, documentation and performance. In other words, they have allowed artists' lives to be seen as one continuous artistic performance in and outside the public gaze. After all, artists are primarily concerned, as Boris Groys correctly suggests, with self-definition.[6] ARIs, with their transdisciplinary ethos and affordable economics, allow artists to innovate, to run wild, to test themselves by creating against market forces and the desire of museum curators to treat artists as a proletariat from which they can draw to illustrate their own propositions.

IS THE ARI
A LABORATORY?

Ironically, contemporary art, dance and theatre have at different times used the term 'laboratory' or 'lab' to indicate that their work or workplace is a site of experimentation and inventiveness, a departure from the gallery or museum.[7] According to the Canadian artist and theorist Sara Diamond, new media artists are more likely—due in part to the transdisciplinary relationship between technologies and art—to work in lab settings; these may or may not bear any relation to a lab in the sciences.[8] However, this term is also used to indicate a more open-ended art-making site, where the process is seen as of greater value than the finished object. This is the opposite of the way the gallery or museum operates, as they place a premium on the finished work—on mastery, authorship, connoisseurship.[9]

At the centre of inventiveness by artists is the complex mutating definition of art. Art is, simply put, everything that serves as art. This means that it includes all the institutions, practices, exhibition spaces and private activities of artists. Therefore, any attempt to precisely define art falls short of the mark. Since Duchamp's time we have known that everything can become art, so the question is what kind of innovation and knowledge does art provide us with?

ARIs have always been laboratories of aesthetic experimentation, social transformation and speculative thinking. This radically contrasts with the classical art world, which is predicated on defining art as a finished object, not as an open-ended process.

It is perhaps ARIs, which are not saddled with the slow-moving bureaucratic structure and public accountability of the museum, that can respond best to the rapid blurring that is occurring due to the impact of new technologies, and in some cases the fusing of art, science and engineering. The notion that ARIs are laboratories and sites of research has taken on a new meaning—and some would say urgency—in light of the research-led culture that has become central to art schools since their amalgamation with universities in Australia, the United Kingdom and across Northern Europe. This has been accompanied by the introduction of higher research degrees, which in some cases have allowed artists to forge links with disciplines outside the art school and the art world.[10]

ARIs are also spaces where the internet and digital media have enhanced inventiveness, in a direct manifestation of the rapid dematerialisation of art and the ramifications of that for today's creative industries and the global economy.[11] This has also meant that the inventiveness that has emerged from ARIs over the last 15 or so years has been based on the image of the artist as social sculptor, someone who engages in social and political critique of the prevailing aesthetic, cultural and institutional orthodoxies and dogmas.

In this fundamental sense, the inventiveness that takes place in ARIs is roughly equivalent to what Hans-Ulrich Obrist describes as taking place with large urban projects when he evokes Mayakovsky's celebrated view that the streets are our paintbrushes and the squares our palettes.[12]

'Modern art' actually emerged in the 19th century. Before then there were only, as Groys reminds us, luxury items such as icons and portraits, and they were basically used and consumed.[13] Art only really emerged when these objects were removed from daily life and placed in museums. They became part of the public realm and not regarded, as they had been previously, as private property.

Therefore, our understanding of art since the advent of the art museum is more inclined to be Hegelian than Kantian in nature. Hegel stressed that art affords us the opportunity to value a certain period of time. This has become central to classical art history and the actual way art functions in our lives today.

The artist is, since Duchamp's era, the artwork, as exemplified by German conceptualist Joseph Beuys and seminal American pop artist Andy Warhol. These two artists specialised in self-representation, which usually means having the ability to control one's image in the media. Art and mass culture have radically shifted since the late 20th century, when most artists were not so concerned with their social representation. Today, because of the internet and social media, everyone has become an artist, a cultural producer. What are the implications here for the role of inventiveness in our world?

ARIs have, historically speaking, focused on contemporary art as 'a world-making activity' that has reflected the complexities of the dynamic, shifting relationship between aesthetics and politics.[14] Thus the inventiveness that has emerged from these spaces can be seen to be primarily part of the discourse of aesthetics, following the definition of artistic practice as a continuous negotiation of 'the social' in society. This has been achieved through non-object art processes and collective participation. In fact, it can be argued that inventiveness in this context stems also from a discursive, self-reflexive relationship between aesthetic and theory. New and emerging artistic practices in ARIs constantly critique the Eurocentric character of art history by explicitly underscoring the diverse North/South, East/West contributions to today's global culture.

Artists who operate in ARIs have tended to value the experimental and hybrid dynamism of contemporary visual practices. Ultimately, this has implied, as Arjun Appadurai put it, that 'the imagination is today a staging ground for action, and not only escape'.[15]

If ARIs have contributed to the idea of inventiveness as an expression of aesthetic cosmopolitanism, that has been because they have critiqued the ideological features of Fordism, neo-liberalism and the culture industry as mass deception.[16] The main question we need to ask, as Gerald Raunig rightly does, is, in the light of the rigorous rejection by Horkheimer and Adorno of the 'economisation of culture', does the present day transformation of the 'culture industry' into the 'creative and cultural industries' promise universal salvation for the majority of the main actors in the cultural field?[17] And where do ARIs sit in this scenario, as artists may have themselves become subjugated to the state and market institutions and forces of the creative industries?

What is interesting in relation to the innovation capital of ARIs is that they have allowed artists to become self-employed cultural entrepreneurs in temporary ephemeral project-based micro-enterprises in the fields of new media, design, graphics and pop-up spaces. However, contemporary thinkers such as Virno, in contrast to Horkheimer and Adorno, regard ARIs as overcoming Fordism and Taylorism.[18] ARIs have certainly played a part in artists becoming a vital part of the more informal, open, non-programmed spaces of post-Fordist production. This is the critical point: since the mid-twentieth-century, the traditional cultural industry entertainment and media enterprises have been essential social subjugation sites, as Raunig points out, and more recently a new form of the individual being subjugated to the control of capital, in what Deleuze and Guattari have called 'mechanic enslavement', has emerged.[19] This can be seen, ironically, in 'the creative industries', as defined by neo-liberal cultural politics and urban development, where we see so-called creatives being released into the spheres of freedom, self-government and independence. Clearly, ARIs, as enabling spaces of inventiveness, creativity and self-reflexivity, are not without their ambiguities and tensions.

Thus, when we discuss the innovative potential of ARIs we need to be mindful of the fact that it is not enough to be a 'creative' working in the new global cultural and informational economy and actively engaged in a rudimentary self-reflexive form of individuation, as Angela McRobbie perceptively argues.[20] To ensure that the economic, the social, the political and the cultural are co-mingled by the artists, cultural producers and others located in these ARIs, they need to battle the flattening anti-egalitarian structures of neo-liberal global homogeneity.[21]

CULTURAL ENTREPRENEURSHIP AND SOCIAL CRITIQUE

This means having the ability to produce ideas, contexts and genres that speak of self-critique and social critique. Contrary to the social engineering of neo-liberal meritocracy—as manifested in, for instance, Tony Blair's England, with the 'young British artist' defined as a cultural entrepreneur or as an 'incubator' in New Labour policy adviser Charles Leadbeater's terms—it is mandatory, McRobbie posits, that the creativity emerging from 'strange and interstitial spaces', one that is fundamentally reliant on a reflexive engagement with communicative networks, images, texts and music, also values 'the productivity of being'.[22] ARIs have the capacity to realise that inventiveness can emerge despite the mass conformity of the neo-liberal global order; 'new productive singularities' (Hardt & Negri) are quite possible when the working body becomes a critical point of intersection with other working bodies.

This important last point was recently underlined in Alexander Provan's insightful article on ARIs.[23] He argued that despite the rhetoric that such spaces are models for establishing a more democratic civil society, it is vital that they are instead seen as places where everyday personal enactments through art have the capacity to 'engage fellow citizens in the difficult work of addressing one another and not some chimerical oppressor'.[24] We need to evaluate ARIs not in terms of what they accomplish in their work, but in terms of what they symbolise in their form.

If ARIs are to be adequately analysed we have to conceptualise them as moral agents, not as spaces of a particular political consensus. This means seeing them in terms of their ethical frameworks, and how those shape our own ethics: the artists working in such institutions have responsibilities that go beyond their usual ones (of ensuring that the 'creative autonomy' of artists is protected) to endeavouring to see to the autonomy of citizens. If ARIs are to matter, it is critical that they speak to us as citizens and not just as artists or cultural producers.

If the ARIs' critical purpose is to mobilise a critique of the institution of art, how do we best evaluate them? Provan asks us, will success be when we are seeing galleries regularly exhibiting 'challenging content', and art fairs, museums and biennials using workshops, discussions, labs, etc?[25] Michael Warner sounds a cautionary note: these sites, which mimic ARIs, do not effectively produce viable public spaces upon which a genuine civil society may be founded. As Warner opines, it is crucial that, when addressing a public, one is producing for that audience a concrete and liveable world.[26]

CURATING: THE ARTIST AS CURATOR AND THE CURATOR AS ARTIST

One reason why ARIs are innovative spaces in the art world's dominant models of exhibition, curation and critical reception is that ARIs are sites where the artist can become the curator. More often than not this is by default, and it may not have been the original intention of the artists who established the ARI. It is, however, the various strategies of intervention, collaboration, participation, critique and networking that, when coupled with a questioning of the role of the global or star curator, biennales, the auction houses, museums and commercial galleries in shaping the content and context of their own exhibitions, are the reasons for this shifting role.

ARIs therefore question the definitional morass of the term 'curate', though they have played their part (historically speaking) in contributing to the opaque specialist language of contemporary curating, with its insistence on explanatory wall texts, the ubiquitous audio recordings, catalogue essays that do not speak to the artworks in an exhibition, and so forth. It is critical to note that one of their innovative features has been to always problematise how museums have shifted from sites of enlightened and inspired iconoclasm to ones of romantic iconophilia, as Groys has recently argued.[27] Thus ARIs have been spaces of innovation because they have always critiqued the way museums have suggested that the curator, not the artist, has turned art into art.

Another important definition of curating is curing. This idea of 'curing' has been described by the French philosopher Derrida as 'pharmacon' curating. Curating 'cures' the image even if it makes it unwell in the process.

Another innovative contribution of the ARIs has been their reliance on making artworks visible as online objects for critical reception and exhibition. This has meant that they have had a drive to explore networks, collaboration and participation in the digital world. These developments have had a profound impact on the role of the independent curator working today in museums, galleries and archives. Thus ARIs have played a very significant role in positioning artworks in mass media through the internet and digital media.

Another of the lasting legacies of ARIs is their role in the emergence of new 'in-between' genres, art events, and documentation. ARIs have always been sites of critical experimentation and curation, and have always been open to continual questioning.

THE PARADOX OF FUNDING

The continually expanding influence of various government funding bodies, at local, state and federal level, requires an ever-greater accountability from galleries and museums. This has a self-censoring impact on curators, directors and the institution in general, and, of course, on artists as they strive to address the competing demands of these various constituents. These constituents often now also include a sponsor or benefactor, and the public relations image of their brand or corporate mission. While some ARIs receive limited public or private funding, in the majority of cases they are self-funding. So, depending where one sits on the political spectrum, ARIs may be seen either as entrepreneurial enterprises in the tradition of the Scottish moral philosopher Adam Smith or as socialist collectives. Perhaps this twilight or paradoxical economic model in fact allows significant inventiveness to take place.

Fundamentally, ARIs are sites of inventiveness because they collectively gather all the vital resources of fellow artists to question and free themselves from the ideological spectacle that is local, state and federal funding for the visual arts, museums, collectors and the creative industries.

ARIs, as public spaces of inventiveness, also need to be constantly critical of the pernicious aesthetic, cultural and ideological consequences of the new audit culture in the arts and in general, as it usually strives to push these public spheres of art-making, curating and documentation into accepting the global consumer culture of branding logos, mega-gentrification and the spectacle.

ARIs also endeavour, as sites of inventiveness, to contest the closing gap between brand management, themed biennales, 'blockbuster' museum exhibitions and the hyperbolic economy of art auction houses, and to mobilise autonomous, politically antagonistic intervention.

ARIs are also spaces of inventiveness because they stress the multiple critical, curatorial and cultural benefits of self-organisation based on real experience, and do not accept the orthodoxy of official culture. They help transform the rules of the bureaucratic and social game into collective political speech.

CONCLUSION

ARIs are places of inventiveness because they insist on deconstructing the control techniques and processes of art bureaucracies, media institutions and critics and reconfiguring them as possible emancipatory strategies to enable us to understand and modulate our everyday lives.

ENDNOTES

1   Henk Slager, *The Pleasure of Research* (Helsinki: Finnish Academy of Fine Arts, 2012), 12.

2   For a history of artist-run initiatives in New York see Lauren Rosati and Mary Anne Staniszewski (eds), *Alternative Histories: New York Art Spaces 1960 to 2010* (Cambridge: Exit Art NY and MIT Press, 2012).

3   For a more elaborate discussion of the recent mergers of art schools with universities and the development of research cultures in schools of art and design, see Brad Buckley and John Conomos (eds.), *Rethinking the Contemporary Art School: The Artist, the PhD, and the Academy* (Halifax: NSCAD University Press, 2010).

4   For a history of Firstdraft and the role of artist-run initiatives in Sydney, see Helen Hyatt-Johnston and Virginia Ross, 'Introduction', in Mark Jackson (ed.), *The Changing Critical Role of Artist Run Initiatives Over the Last Decade* (Sydney: Firstdraft, 1995), 19–29.

5   Oscar Wilde, 'Preface' in *The Picture of Dorian Gray* (New York: Simon & Schuster, 2005[1891]), 4.

6   Boris Groys, 'What is Art/Everyone has to be an Artist/After the End of Mass Culture/Egypt as a Quick Run', in Ingo Niermann (ed.), *The Future of Art: A Manual* (Berlin: Sternberg Press, 2011), 221.

7   Beryl Graham and Sarah Cook, *Rethinking Curating: Art after New Media* (Cambridge: MIT Press, 2010), 234–36.

8   Graham and Cook, *Rethinking Curating*, 237.

9   Hans-Ulrich Obrist, 'Expanded Curator/The Art World As Salon/The Found Pyramid', in Niermann, 2011, 138.

10 On the role of PhDs in schools of art and design, see Buckley and Conomos *Rethinking the Contemporary Art School*.

11 The term 'the dematerialisation of art' in the 20th century was first used by Lucy Lippard: see Obrist, 2011, 137.

12 Cited in Obrist, 2011.

13 See Groys, 'What is Art', 219.

14 Nikos Papastergiadis, *Cosmopolitanism and Culture* (Cambridge: Polity Press, Cambridge, 2012).

15 Quoted in Papastergiadis *Cosmopolitanism*, 95.

16 See Gerald Raunig, 'Creative Industries as Mass Deception', in Marc James Léger, *Culture and Contestation in the New Century* (Bristol: Intellect Press, 2011) 94–104.

17 Raunig, 'Creative Industries', 95.

18 Raunig, 'Creative Industries', 98–100.

19 Raunig, 'Creative Industries', 100.

20 Cf. Angela McRobbie, 'Everyone is Creative: Artists as Pioneers of the New Economy?', in Léger, 2011, 79–92.

21 McRobbie, 'Everyone is Creative'.

22 On Charles Leadbeater's ideas on the arts and culture and the new informational economy for Tony Blair's New Labour government in the 2000s and Hardt and Negri's concept of the emergence of a new mode of becoming human with this new information and communications technologies, see, respectively, McRobbie, 2011, passim, and 90–91.

23 Alexander Provan, 'All for One', *Frieze*, no. 153, March 2013, 117.

24 Provan, 'All for One'.

25 Provan, 'All for One'.

26 Provan, 'All for One'.

27 Boris Groys, 'On the Curatorship', in *Art Power*, ed. Boris Groys (Cambridge MA: MIT Press, 2008), 43–51.

ICAN: REINVENTING THE AUTONOMY
OF THE ARTIST-RUN INITIATIVE

ALEX GAWRONSKI

The Institute of Contemporary Art Newtown, or ICAN, is an artist-run initiative (ARI) operating in the inner suburbs of Sydney, Australia. The project was instigated in late 2007 by myself, fellow artist Carla Cescon, and gallerist, curator and artist Scott Donovan. Sydney artist Stephen Birch (1961–2007) was initially a driving force behind the project. Over its approximately five years of operation, ICAN has exhibited diverse and inter-disciplinary contemporary work by artists based in Sydney, across Australia and internationally. Overall, ICAN was conceived not so much as a simple gallery aimed primarily at exhibiting discrete art objects, but as an experimental space functioning as a type of critical laboratory. Such an approach has fostered a particular climate of inventiveness, drawing together networks of artists, many of who have significant shared histories. Such inventiveness has also been characterised by a broadly anti-commercial stance, which is to say that ICAN has consciously sought to emphasise the value of art as a form of experimentation over its instant reduction to a series of commodities made available to a purchasing public. This outlook has sought, by example, to place the ARI beyond the ambit of others whose primary self-designated, and paradoxical,[1] motive seems to be the generation of profit through sales, buoyed by the attendant gloss that symbolic proximity to markets presumes in an age of globalised capital. ICAN's emphasis has been on redefining art production as resulting in something other than the presentation of singular commodities.[2] ICAN has hoped to show that ARIs can become a primary site for invention by fundamentally reframing contemporary art as a visual form of non-instrumentalist thinking. Such thinking about art has led ICAN in a number of interrelated directions that have included: the production of ambitious, site-specific work by mid-career and established artists; the realisation of thematically curated group exhibitions bringing together artists whose interests either strategically collide or coalesce the facilitation of cross-platform exchanges with artists and artist groups based locally, interstate and internationally; and, the pursuit of critical fictions[3] that implicitly target the fictive dimension[4] of all institutions. The last approach has also periodically included ICAN's deliberate production of disinformation in the guise of occasional ads for the gallery placed in art journals.[5] Indeed, the self-nomination of ICAN as an Institute of Contemporary Art critically and playfully calls into question the automatically assumed unassailable legitimacy of genuine institutions, including those encompassed by the woefully de-politicised agendas of contemporary party politics.[6] While ICAN has no party-political aspirations, nor is it an activist space, its politicised reinvention of the ARI—itself a highly problematic term[7]—as a significant site of at least partial autonomy, allows invention to multiply within the open parameters it has set for itself. The best way to address the way in which invention occurs within ICAN would be to discuss in some detail a number of its interconnected projects.

An important aspect of ICAN's ongoing focus has been the presentation of experimental, site-specific or generally non-commercially orientated work by mid-career and established artists. This gesture is also expressly political in that it consciously aims to redraw attention to the ARI, not as a 'training ground' for so-called emerging artists,[8] but as a key venue for the facilitation, development and presentation of work by contemporaries with significant histories. Paul Saint is an established Sydney-based artist with an exhibition history spanning at least 20 years. Saint's more recent exhibitions have included a survey show at Sydney's Artspace Visual Arts Centre that was accompanied by a monograph launched at ICAN in 2012. Saint's work encompasses sculpture, photography and digital media. The artist has shown twice at ICAN in 2008 and 2012. In 2008, Saint exhibited *Rule*. The show consisted of three large mirrors distributed across three walls of the gallery. The mirrors were lined like the pages of a textbook and bordered by industrial steel rulers glued directly to their surfaces. The installation functioned not only to question the traditional illusionism of representational space but to further multiply the artwork's frames of reference. These mirror works achieved this telescoping effect by simultaneously reflecting each other, the interior of the gallery, the street outside as well as any visitors to the space. While the works were serialised and discrete, their inherently spatialised relationship to each other meant they were ultimately irreducible to singular removable units.

Similarly site-dependent was an installation by another established Sydney artist, Wade Marynowsky, who also exhibited at ICAN in 2008. Marynowsky's exhibition *The Discreet Charm of the Bourgeoisie Robot*, presented a remotely interactive robot in a black Victorian-style dress, sewn by the artist's mother, flanked on either side by antique-looking speaker horns. Marynowsky's piece functioned on multiple levels: it implicitly questioned the physical centrality of the artist by replacing him with an avatar; it extended the space of the gallery virtually, as the artist, under the alias 'Boris', spoke to the audience from afar via a networked computer interface; it pushed the disciplinary boundaries of visual arts and 'sculpture' into the technical fields of computer programming and robotics. Marynowsky's domesticated robot inhabited the gallery as though it were an alien environment, in turn reinventing the space via the inescapable uncanniness of its wireless presence.

Lastly, and evincing a related attitude of spatial colonisation, established practitioner Sarah Goffman, in conjunction with her partner Peter Jackson, presented *Fatty and Slender; the Hanging Man's House* in 2010. For this large-scale installation, Goffman translated her recent experiences as an artist-in residence in Tokyo, by reconfiguring ICAN according to a Japanese sense of interior space. However, Goffman's exhibition was by no means a mere touristic exercise in exoticist re-creation. Instead, the installation appeared as an urban accretion of objects and materials drawn from the artist's recent memories of other, very distinct, environments. In this instance, deploying a distinctly 'maximalist'[9] approach, almost every inch of ICAN was transformed to suggest a type of ad-hoc quasi-private dwelling; walls were covered in cardboard, makeshift tables were arranged with Jackson's semi-surreal carvings, mats covered the floors, handwritten notes lined one of the windows and stacks of discarded clothing were piled up to suggest a casual totem. Having exhibited many accretive works of this kind over a long period, Goffman's complete reinvention of the space of the gallery, spoke of the transient aspects of global art production, and of the itinerant artist's repeated attempts to reinsert a modicum of homelessness into comparatively alien situations.

Figure 4.1: *Welcome to the desert of the REAL (or a place for new songs)*, Brad Buckley, 2008, paint, vinyl text, dimensions variable, ICAN, Sydney. Photo by Rowan Conroy.

Figure 4.2: *EIDIA Occupy's ICAN*, 2012. ICAN, Sydney. Photo by Alex Gawronski and ICAN.

Viewed together, these works are but a small sample of experimental exhibitions by established artists staged at ICAN, each through their ambition, complexity and the sophistication of their references, succeeded differently in questioning common assumptions of the ARI as simply a training school for 'arts professionals'.

Considering this emphasis on individual experience and multi-skilled expertise, ICAN has, from its inception, defined itself predominantly as an independent autonomous entity. That is to say, the activities of the gallery have always been directed towards the nuanced exploration of concepts that are not overtly beholden to external funding or other commercial criteria. Given this, the many curated projects ICAN has staged during the course of its operation have drawn upon ensembles of artists to illuminate specific themes in ways that highlight the gallery-container as a locus of, and for, invention. Good examples of this approach, arrived at collectively, were the group exhibitions *Headland* and *The Australian Pavilion; Swallow it Dog! (after Francisco Goya)*. *Headland* was initiated in 2008 by Scott Donovan in conjunction with Caleb Kelly, a lecturer and theorist of contemporary sound art.[10] The exhibition brought together a diverse range of contemporary local, interstate and international experimental sound artists whose works were available to listeners for the duration of the show. This emphasis on sound, emerging partially out of Kelly's long-running independent presentation of audio works on radio and live under the banner of 'Impermanent Audio', served to spatially reinvent the space of the gallery as a temporal, ephemeral, audio-sonic environment. Via these means, listener/visitors were transported beyond an automatic assumption of the inherent visuality of art galleries. That said, part of the curatorial brief of this exhibition was that each audio work had to be accompanied by a related visual artifact. In this way, the at-times hermetic, consciously anti-visual attitude of sound artists, was tested and complicated by asking all participants to somehow *visualise* an audio context. The fact that such a relationship, between sound and image, will never be merely illustrative only assisted in forcing both sound artists working visually, and visual artists working with sound, to inventively reconsider the trans-disciplinary connection between these fields. The outcome of the exhibition was exceptionally varied and inventive as result of sound and visual art being curatorially encouraged to productively 'contaminate' each other's disciplinary parameters.

Of a different calibre and intent was ICAN's curated exhibition, *The Australian Pavilion 2010; Swallow it Dog! (after Francisco Goya)*.[11] Staged in 2010 to coincide with that year's Biennale of Sydney, the show was the second in what has become a regular series of independent responses to aspects of the Biennale. Again, the approach of this show sought to prefigure the ARI as an autonomous entity with its own processes and ultimately in control of its conceptual and curatorial boundaries. However, this continuing emphasis on autonomy does not naively presume a totalised outside to the 'insider' worlds of institutional or commercial art. In fact, it is precisely in relation to these spheres that many of ICAN's curated exhibitions have taken place. At times the gallery has staged events, launches or openings for sympathetic state-funded venues or drawn from a pool of well-established contemporary artists with notable commercial profiles.[12] Thus, the autonomy of ICAN is necessarily partial and indeed functions more forcefully through recognising its embedding in a much wider global contemporary art scene. Nonetheless, *The Australian Pavilion 2010; Swallow it Dog! (after Francisco Goya)*, like most of the unofficial Biennale exhibitions ICAN has presented, sought to critically (and humorously) address the monolithic dimensions of the Biennale of Sydney apparatus. The 2010 Biennale of Sydney, 'THE BEAUTY OF DISTANCE: Songs of Survival in Precarious Age',[13] although alluding somewhat hyperbolically to global conflict economic and ecological catastrophe, tended nevertheless to purvey like many contemporary biennales a rather diffuse and spectacle-driven agenda.[14] Consciously invoking its self-appointed institutional status and satirically mimicking the 2010 Biennale of Sydney's exaggeratedly attenuated title, here ICAN invented its own micro-biennale that aimed to obliquely suggest some of the more troubling, 'unacceptable' phenomena apparent on the contemporary socio-political horizon. Furthermore, the artists involved in this show intentionally included those working as academics interested in the pedagogical ramifications of art practice. Their work referred to often unspoken institutional realities of the exhibition context: the New Zealand born, Canadian conceptual artist and academic Bruce Barber produced a textual inventory of the many possible permutations of the contemporary terrorist bomber; local artist and academic Andrew Hurle presented a perfect imitation of a Taschen coffee-table art book called *100 Nazi Girlfriends* while the Viennese based collective 0-gms, included in the 'real' 2010 Biennale of Sydney, exhibited a video work. The first of these works ironically alluded to the sheer hysteria propelling anti-terrorist rhetoric where everyone, including the artist, is a potential suspect; the second commented, quite literally, on the inherent fetishistic dimension of the 'art book'; the third, in which the artist-speaker pedantically reads out a list of all the global biennales currently operating while making comments about which were worth participating in, alluded simultaneously to the paradoxical proliferation and structural exclusivity of the biennale format. In each instance, the artist presented work of 'biennale quality' yet whose content was somehow too cannily aware of the superficially differentiated, though in reality quite generic, art of the regular doyens of the global biennale circuit. Here, ICAN invented a platform from which those slyly, even 'accidentally', dissenting voices not evident in the 2010 Biennale of Sydney itself could be heard. Considered together, the exhibitions *Headland* and *The Australian Pavilion 2010; Swallow it Dog! (after Francisco Goya)* evidenced the capacity of ARIs to independently engage across technically unrelated fields and in response to organisations whose institutional prioritisation frequently goes unquestioned.

In a similar spirit to its curated exhibitions, ICAN has also co-facilitated a number of exchanges in Australia and overseas. Of these, two are worthy of particular mention as both evinced a distinctly networked approach, building upon and expanding ICAN's contacts as well as its metamorphic capacity for ongoing curatorial invention. The first exhibition was an exchange between the ICAN directorship and that of Ocular Lab (1997–2009), an ARI located in West Brunswick, Melbourne. The second exchange was facilitated between ICAN and EIDIA House (Plato's Cave), Brooklyn, New York. Ocular Lab was a collective[15] known for its embrace of inter-venue dialogue and a history of off-site hostings of exhibitions and events. In 2009, Ocular Lab staged an exhibition at ICAN titled *The Gift*. The show came about primarily through contact with Elvis Richardson, an artist who had resided in Sydney for many years and had previously been involved with artist-run projects like Firstdraft and Elastic.[16] Richardson, during her time in Melbourne, had connected with the significant group of contemporary artists comprising the Ocular Lab directorship. ICAN joined this network through Richardson to realise a two-part project. Ocular Lab's exhibition at ICAN encompassed video, conceptual photography, objects and the residue of a street postering action. Indeed, the title of the show paraphrased not only theoretical anthropologist Marcel Mauss' famous text of the same name,[17] but, perhaps more acutely, French philosopher Georges Bataille's extension of Mauss' research into a theory of a counter-economy of jouissance and excess.[18] Thus, Ocular Lab in this case prefigured the exhibition as a model of invented—that is to say, non-functional—exchange wherein the value of the exchange remained open to debate according to the presumptions of the audience. Moreover, the potential generosity of the exhibition as a specific cultural manifestation was highlighted, albeit without recourse to fey moralism. At the same time, the intrinsic ambiguity of *The Gift* as theorised by Mauss and Bataille was simultaneously and paradoxically, foregrounded. Responding expressly to this underlying structural ambiguity, ICAN's show at Ocular Lab, *Quinto Sesto; Fighting for Peace: Part 2*, took the form of an array of found materials—notebooks, posters, a manifesto, discarded clothing, black and white photographic portraits, a hand-painted banner—purportedly representing an archive of a little-known and unsuccessful local terrorist/art organisation called Quinto Sesto (1975–76). Given its controversial, even spurious, content, viewers were asked to consider what the relationship between art practice and actual political subversion, as well as its consequences, might be. Such content was especially pertinent in a contemporary context given that at no time since the 1970s had the term 'terrorism' come to haunt the imagination with such ubiquitous vengeance. Even more unsettling to viewers of this archive was the fact that they were perpetually asked to question its factual veracity, which was never explained. Rather than just as an offering of isolated commodities, the exhibition format was reconsidered as the ground for testing contemporary art's greater relationship to truth and to particularly urgent socio-political realities.

Equally inventive in scope was the exchange between ICAN and EIDIA House (Plato's Cave). EIDIA House is an independent project operating transversally across spaces while its linked venue, Plato's Cave, is a project space connected to the studio of its founders Paul Lamarre and Melissa P. Wolf. EIDIA House is an endeavour especially inventive in its generation of its own institutional alternatives.[19] The exchange between EIDIA House and ICAN was initially conceived out of Lamarre and Wolf's participation in the 2011 artist in residence program at Sydney College of the Arts, the University of Sydney. EIDIA House has consistently linked its projects to the contemporary political sphere.[20] The exhibition they presented at ICAN in March 2012, *EIDIA Occupy's ICAN*, was no exception.[21] In this instance, EIDIA showed a series of Lamarre's colour documentary photographs of the Occupy! protests mounted all over New York City and globally at that time. These photographs were accompanied by the detritus of Occupy! events— posters, flyers, invitations, hand-written notices— and a video that captured the multi-venue, simultaneously carnivalesque and enraged, chaos of those protests. Notable about this exhibition, however, was its avoidance of a readily transparent political agenda; in no way did the show degenerate into a simplistic propagandistic statement by its authors. Instead, a complex ambiguity underscored the portrayal of these often raucous gatherings. As a result, the exhibition assumed paradoxical museological and activist identities.[22] In a related but wholly different vein, in April 2012 ICAN presented *ICAN Occupy's EIDIA* at Plato's Cave. Suggestive of its name, Plato's Cave is physically located under street level in Williamsburg, New York. Given this unusual siting, the ICAN directors responded broadly to the implications for art of the term 'underground'. Of course, this term is culturally loaded and has been variously disabused of its original overtones of radicality and subversion. Nonetheless, the sheer secretiveness of the Plato's Cave venue granted it subversive connotations that were aligned with ICAN's re-configuration of the tiled subterranean space as a place of inverted values. Consequently, the symbolism of execution, objectification and political mud-slinging that permeated the show[23] could be read as a rejoinder to the unsavoury underside of art's historical ties to the Enlightenment project. These dual, interconnected but distinct exhibitions, *EIDIA Occupy's ICAN* and *ICAN Occupy's EIDIA*, challenged the automatic assumption of the gallery as a space for the passive contemplation of art. Each reinvented the ARI as an arena from which contemporary socio-political realities could be addressed and in forms that may or may not have even been perceived as art per-se.

This deliberate and productive confusion of disciplinary boundaries pursued in the aforementioned exhibitions has extended into a number of related ICAN shows that have sought to blur the lines between museology and fiction. This general attitude has also been pursued in the promotional material ICAN has produced over its lifespan. A good example of ICAN's mixing of quasi-museological and essentially fictional content was the 2012 exhibition *We Are all Transistors*, staged by the gallery at the Aratoi-Wairarapa Museum of Art and History in Masterton, New Zealand. The realist basis of this show emerged out of director Scott Donovan's intimate knowledge of Masterton and of the work of the local modernist architect Roger Walker. *We Are all Transistors* concentrated particularly on Walker's Centrepoint complex erected in 1974 and demolished not long after. The curious animosity directed by locals at Walker's structure for its strangeness and provincially

Figure 4.3: *Fatty and Slender; the Hanging Man's House*, Sarah Goffman, 2010, ICAN, Sydney. Photos: Alex Gawronski and ICAN.

Figure 4.4: *Corpse*, Quinto Sesto, 2012. In the exhibition: 'Moving Forward to the End',
curated by Iakovos Amperidis, 55 Sydenham Rd Marrickville, NSW. Photo: Iakovos Amperidis.

challenging 'internationalist' sympathies became, in ICAN's exhibition, a conduit for a greater rumination on failed utopias. Indeed, it was precisely the formal and spatial inventiveness of Walker's buildings that sealed the fate of their ultimate demolition. ICAN deployed this paradox to invent a number of quasi-fictional reference points that explored this very localised instance of cultural chauvinism beyond the limits of its initial affect. Therefore, alongside documentary photographs of the building taken during and after its completion, a swathe of the architect's original working drawings, the surviving conical summit of the actual structure[24] and paintings by the architect's son, the artist Jake Walker, of his father's buildings were other puzzling non-archival artifacts, sculptures and video. Light fittings that appeared to belong to the period, drooped pointlessly onto the gallery floor; a monochromatic hand-made bust of Karl Marx was perched atop a pile of concrete pavers suggesting the political origins of failed utopian social constructions; an installation entitled the *Dresden Lounge,* incorporated a video of rocks being thrown against the ground to the perpetually agonised strains of contemporary Polish composer Krzysztof Penderecki's stridently apocalyptic *Threnody for the Victims of Hiroshima* (1960). Embracing irony, this installation spoke especially of the precariousness of all utopian projects while by no means discounting attempts at their realisation, including via the pursuit of the partial utopia of self-directed ARIs. Overall, *We Are all Transistors* in its mixing of concrete and fictional artifacts, as well as via its specific harnessing of the significant work of a local creative figure, intended to re-propose museology as a variety of invention dependent as much on knowledge as on improvisation and subjective free association.

With this in mind, the promotional material ICAN has produced, while by no means voluminous, has aimed to interrogate the very notion of advertising as a means of merely relaying useful information or strategically enticing new audiences. Alternatively, ICAN's ads have all aimed to inventively address the art-advertising context to expose the unspoken institutional and commercial imperatives driving such ads. ICAN's earliest ad, a still-life staged and photographed by its directors, featured a bottle of red wine, wine glasses, a wedge of Swiss cheese, a picnic basket containing a fresh fish head, and a travel bag out of whose front pocket peeked a touring map of France's famous Cote d'Azur. However, behind this intentionally art-clichéd image, hung a provocative text work by Stephen Birch[25] that instantly counteracted the faux gentility of the photographic still life through explicitly evoking French Dadaist Marcel Duchamp's career-long interest in the erotics of art. In another ICAN ad, a ventriloquist dummy and his operator grinned ghoulishly above a text asking 'Are you trying to undermine me?' In both cases, beyond utilitarian references to ICAN's location and opening hours, no information about the gallery or its program were provided. These ads implicitly questioned the instrumentalisation of art through advertising in an artistic climate impacted heavily by often restrictive funding criteria expecting concrete commercial outcomes.

Over the five years of its existence, the Institute of Contemporary Art Newtown has aimed to re-propose the ARI as an independent ARI through which inventive, spatially and temporally exploratory contemporary art might be realised and encouraged. ICAN has expressed through its multiple activities undertaken in Sydney, interstate and internationally a commitment to contemporary art as a vital tool for critical inquiry and speculative thinking. Once again, this is often in distinct contradiction of local and global art-world trends increasingly beholden to the domineering dictates of neo-liberal markets and a related general culture of high spectacle. The continual reinvention of ICAN as a laboratory for the pursuit of active critical thinking about art from a local perspective, consciously extending outwards, is evidenced on a number of interrelated levels. It is apparent in ICAN's ongoing fostering of experimental work by mid-career and established artists, it is evident in the gallery's conceptual staging of inventive curatorial enterprises that deliberately engage artists in either sympathetic or tense relationships, it is clear in its facilitation of local and international exchanges with spaces of similar affiliation and it is equally manifest in its deployment of critical fictions to self-reflexively address both ICAN's own activities and those occurring within the greater art world of which it is a part. Crucially, each of these orientations is connected to the growing network of artists, independent curators, writers and academics that surround and define the space. Finally, ICAN's multi-tiered approaches demonstrate that independent creative activity generated by practising contemporary artists can affect a wider climate of invention. It can achieve this through foregrounding knowledge, not primarily as a commodity, but as a free field of conceptual recombination out of which sedimented institutional truths may be continually questioned.

ENDNOTES

1   Such an attitude is paradoxical as the historical raison d'être behind the formation of ARIs has been the pursuit of *alternatives* to existing commercial and institutional art gallery models.

2   Over the past three years there have been practical exceptions to this rule as ICAN has staged fundraising auctions to which many established artists have donated works. The money raised by these occasional auctions, which are not true exhibitions, has aided the gallery in its independent pursuit of experimental practices.

3   The 'critical fictions' referred to here are exhibitions and art works that self-consciously take into consideration the exhibition contexts in which they are shown. Thus, rather than imagining art merely as the packaging of essentially fictive narratives for passive consumption by audiences, critical fictions assume that the gallery context itself is a partial fiction that can be self-consciously activated as art content. In this sense, critical fictions extend aspects of the wider historical project of 'institutional critique', a tendency emerging out of and aligned with conceptual art in the 1960s.

4   This point relates to the former in recognising that every institution is founded on the fiction of its own authority. Whilst such authority—political, educational, scientific, technological—is undoubted, it is also always questionable. To take but one example, the global financial crisis merely exemplifies the fiction of the authority of the neoliberal free market that is repeatedly represented in the media and elsewhere as both controllable and reliably in control of economic outcomes. When such a representation is proven to be deeply fallible, the fictive dimension of the institution is revealed.

5   Primarily for the journal *Column* published by Artspace Visual Arts Centre, Sydney (vols. 2, 3, 5) as well for Sydney's *Runway* magazine, published by the Invisible inc. (Issue 11, 'Conversation').

6   It has been widely argued by contemporary theorists as different as the French philosophers Alain Badiou, Jacques Rancière and Chantal Mouffe, as well as Slovenian philosopher Slavoj Žižek, that the party politics of contemporary Western democracies are predominantly administrational and managerial and furthermore, that by avoiding the transparent pursuit of warring ideologies, they have effectively evacuated and therefore left empty the 'agonistic' space of genuine politics.

7   The term ARI is problematic as, unlike a more encompassing and perhaps poetic term like Artist Space, it is generically used primarily in a funding context to refer to spaces primarily exhibiting emerging artists. Emerging artists are considered from a funding perspective to be artists in the first five years of their practice. Thus, locally today the terms ARI and 'emerging artist' are implicitly regarded as coextensive. This is particularly problematic for artist-run organisations like ICAN who do not principally exhibit emerging artists and view artist-run activity as an exercise in the practice of a politicised autonomy. See Alex Gawronski, 'Artist Run = Emerging—a Bad Equation', *Art Monthly Australia*, Issue 257, March, 2013 and 'Out of the Past: Beyond the Four Fundamental Fallacies of Artist Run Initiatives', We Are Here Symposium, NAVA, Sydney, September 2011.

8   See previous footnote.

9   'Maximalist' art would be the antithesis of the minimalist art of the 1960s, famous for its exacting reduction of forms.

10 See Caleb Kelly, *Cracked Media* (Cambridge: MIT press, 2009).

11 Goya's print in question, 'Swallow it, dog' (Tragala perro—1799) was part of his famous suite of etchings Los Caprichos (The Caprices) and depicts ecclesiastics of the Spanish Inquisition attempting to force an enema syringe on a victim. The ICAN exhibition simply updates Goya's anti-institutional critique.

12 Such artists, to name a few, would include Hany Armanious, Stephen Birch, Julian Dashper, Mikala Dwyer, Rosemary Laing, Lindy Lee, Mike Parr, Pipilotti Rist, Tim Silver and Jenny Watson.

13 The 17th Biennale of Sydney (2010), *The Beauty of Distance; Songs of Survival in a Precarious Age*, curated by David Elliott.

14 See Alex Gawronski, 'Show me Good Art: Biennales and Artist Directed Culture', *Broadsheet*, ed. Alan Cruickshank, Contemporary Art Space of South Australia (CACSA), vol. 39, no.2, June, July, August.

15 Ocular Lab's members included the artists Sandra Bridie, Damiano Bertoli, Raafat Ishak, Sean Loughrey, Jonathan Luker, Sally Mannall, Tom Nicholson and Elvis Richardson.

16 Richardson went on to co-found with fellow artist Claire Lambe Death Be Kind a death-themed Artist Space in Melbourne (2009–2011).

17 See Marcel Mauss, *The Gift: Forms and Functions of Exchange in Archaic Societies* (London: Cohen & West, London, 1969).

18 Bataille's anti-capitalist counter-economy recognised that missing from capitalism's emphasis on exchange-value as well as from Marx's critique of it, was an unaccountable libidinal excess that could not be usefully 'put to work'. The concept of The Gift for Bataille engages with this remainder, suggesting that a gift economy is distinguished by its refusal to accept exchangeable equivalences. The Gift is notable because it exceeds its capacity to be merely equaled by the receiver: the Gift is always a challenge to the other. See Georges Bataille, *The Accursed Share: An Essay on General Economy* (New York: Zone Books, 1988–1991).

19 Especially pertinent in this regard is Lamarre and Wolf's ongoing project *The Deconsumptionists, Art as Archive* (2006–) which is a full size semi-trailer with the words 'The Deconsumptionists' stenciled in red on its sides. The trailer functions as a mobile archival gallery of the artists' many collaborations, it includes labeled boxes, objects, slide archives and video works. The artists have used this mobile platform to exhibit and talk about their archive in various locations. In its guise as a storage container, it also serves more philosophically as a rumination on the contemporary effects of capitalism and consumption. See www.eidia.com/deconsumptionists/ Accessed 2 July 2013.

20   Given this, Lamarre and Wolf have written an EIDIA Manifesto and produced work
      addressing everything from sexual politics (the Chelsea Tapes—1993) to the
      institutional politics of being an artist (We Apologise—2004). In the latter work,
      the artists satirically apologise for everything they have ever done, including
      producing art. See www.eidia.com. Accessed 2 July 2013.

21   The title is deliberately grammatically incorrect in order to allude directly
      to the Occupy! movement.

22   The latter were so troublesome to the owner of the adjacent café that he suspected
      the gallery had actually become an activist drop-in centre.

23   For this show I produced a life-size guillotine blade that sat just inside one of the
      two studio windows, Carla Cescon made a forlorn and uncomfortably adult-looking
      doll, while Scott Donovan showed a video in which he performed as a dancing potter:
      working on a potter's wheel, the artist, wearing white, gradually spattered the
      room in which he performed with clay and slurry. See icanart.wordpress.com/
      i-c-a-n-offsite-projects/4-i-c-a-n-occupys-eidia/. Accessed 2 July 2013.

24   This was recovered by a Masterton resident who spied it in the front yard of the local
      auto-wreckers inscribed with the handwritten words, 'Apollo (sic) II Space Junk'.

25   The work reads 'Lay off Duchamp you Cunts' and punningly refers to Duchamp's
      final work, the large-scale installation Étant donnés (Given: 1 The Waterfall, 2. The
      Illuminating Gas) (1946–1966). Duchamp's work invited audiences to peer through a
      crack in an antique door installed in the gallery wall, upon which they would glimpse
      the sculpted figure of a naked women lying spread-eagled on a bed of twigs holding
      a gas lamp in front of an artificial waterfall.

5

INVENTION AT THE
CITY
SCALE
[ARCHITECTURAL AND
SPATIAL CONFIGURATIONS]

FIT TO BURST: BODIES, ORGANS
AND COMPLEX CORPOREALITIES

CHRIS L. SMITH

ENTANGLED: COMPLEX BODIES
AND SENSATE MACHINES

DAGMAR REINHARDT
LIAN LOKE

FIT TO BURST: BODIES, ORGANS AND COMPLEX CORPOREALITIES

CHRIS L. SMITH

The aim of this chapter is to explore the wonderful speculative productions of Dagmar Reinhardt and Lian Loke, and in particular their work *Black Spring* for the 2012 *Biome: Digital Interdisciplinations* exhibition held at the Tin Sheds Gallery at the University of Sydney. What I am interested in here is the ecology of ideas: of the logics, disciplines, controls, liberations, and forces that come to breathe life into creative work. In order to *think* the creative productions of Reinhardt and Loke, I will draw upon the work of the philosopher Gilles Deleuze and his ideas related to speaking, eating, and the mouth.

I approach the work of Deleuze (and his collaborative work with Félix Guattari) as an architect and a lecturer in architecture. What this means, I think, is that I approach it with an eye trained on its spatiality—or, more accurately, I approach it with an eye trained on its 'spatial determinations'. Thus when I approach a text such as Deleuze's *The Logic of Sense* (1990), which is a focus in this chapter, what I am doing is thinking through its space.[1] As a text, *The Logic of Sense* wraps itself in long dualistic lists— eat/speak, depth/surface, interiority/exteriority—the type of lists that later works of Deleuze endeavour to undermine. The text remains fascinating, however, not only in terms of charting the movement of concepts through the works of Deleuze, but it is particularly rousing when thought through in terms of the spatial determination of concepts and the manner by which innovative creative works might be thought.

A distinction Deleuze makes between assemblages is very often expressed in terms of different kinds of *spatial determination*. It is an awkward task that Deleuze sets himself in a reliance upon a spatial formation to describe non-representational concepts. We can see this played out in the persistent engagement of metaphor (of wasps and orchids) whilst insisting that the use of language is not metaphoric but conceptual.[2]

This conflict of metaphor and concept is mediated by the use of performative over referential descriptions, and the work of Deleuze becomes reliant on non-subjectivist terms in referring to movements, abstract lines, and various kinds of processes. In the case of 'space' Deleuze tends to deal with the formation and the life of space rather than a unitary character of space itself. This phrasing tends to make the work of Deleuze both seductive and valuable to architects, artists and those who deal with creative practices. It is seductive in that to read a text like *The Logic of Sense,* as an architectural theorist, is to read a text that may appear to be already written in the languages of architecture. It is valuable, however, when one finds that these superficial resemblances indicate connections of a much more complex and problematising mode. The present lament amongst architectural theorists engaged with the work of Deleuze and looking upon its 'use' in architecture is similar to that expressed by the philosopher Jean-Jacques Lecercle when recounting his story of a yuppie reading Deleuze and Guattari's last collaborative text *What is Philosophy*? on the Paris Metro.[3] It is the lament that the work of Deleuze may be forcibly read in such a way as to reinscribe the very characteristics of thought (of space, of ideology) that it seeks to undermine. In architectural theory it must be continually reminded that descriptors such as *depth* and *surface*, or *striated* and *smooth* remain 'spatial determinations' rather than 'spatial categorisations'. For Deleuze, space is the production of spatiality, and very much already an architectural endeavour.[4]

Reinhardt and Loke's project *Black Spring* is best summarised by Reinhardt and Loke themselves as '[t]he design and implementation of hybrid environments—or sensate machines—[that] can be conceptualised as providing new environments and stimuli with which to re-pattern the brain.'[5] The concentration on the design of an environment is not coincidental. Reinhardt is an academic working in the Faculty of Architecture at the University of Sydney. She trained as an architect and continues to practice architecture. Reinhardt's doctoral thesis explored the relation between fashion and architecture in respect to 'latent formations'. Loke is also an academic working in the Faculty of Architecture at the University of Sydney. Loke has degrees in science and engineering. She has also studied fashion design. Loke's doctorate was in the area of interaction design. Reinhardt and Loke are singularly and collectively interdisciplinary. It is also worth noting that beyond Reinhardt and Loke, a project such as *Black Spring* requires a multitude of students, curators and expert technicians in scripting, parametrics, 3D modelling, and digital fabrication.[6]

Despite their architectural leanings, Reinhardt and Loke are not interested in producing installations or backdrops but, rather, what they call 'sensate machines'. They too insist that their words are not metaphorical but conceptual. The idea of the sensate machine is one that Reinhardt and Loke pursue through all their work. It is a conceptual label that their productions circulate about, rather than a label that seeks to articulate the field of the endeavour itself. In this sense the words engaged by Reinhardt and Loke are almost as performative as the creative output. The very idea of the sensate machine is one that unites the corporeality of the sensate and the incorporeality of the machine. The project involves multiple bodies responding, dancing, with one another. For Reinhardt and Loke, bodies 'may not be conceived as fixed and separate entities but instead as constantly changing, through an infinite and complex system of processes occurring in and outside of the bodies; and their relationships with an other'.[7]

This sensate quality in the work of Reinhardt and Loke involves at once a responsiveness of the work to *a* body but also that the work itself has a particularised manner of responding, that another must orientate itself to. That is, a work such as *Black Spring* is not a work that may be objectified *on its own* but is activated by an engagement with another body, in this case, a dancer. The dancer responds as much as the work. In this sense the work has two particular mechanisms in play. On the one hand the work is about a type of control: 'if you do this, I do that'. On the other hand, the work is about a type of self-generating or endogenous force: 'I will do it my way'. It is this dual logic that I seek to explore in terms of Deleuze's ideas of the mouth. In *Black Spring* one body frees the other to express, much as Deleuze writes in *The Logic of Sense* of the relation between the mouth and the brain:

> The mouth is not only a superficial oral zone but also an organ of depths, the mouth-anus, the cesspool introjecting and projecting every morsel. The brain is not only a corporeal organ but also the inductor of another invisible, incorporeal, and metaphysical surface on which all events are inscribed and symbolised. Between this mouth and the brain everything occurs, hesitates, and gets its orientation. Only the victory of the brain, if it takes place, frees the mouth to speak, frees it from excremental food and withdrawn voices, and nourishes it with every possible word.[8]

It would be fair to say that architects have always occupied the mouth. Their stock and trade are eating and speaking and an endless negotiation of surfaces and depths, more often than not in a blind search for good *taste*. Mouths, however, are not only about taste: when John Ruskin, the pre-eminent English art and architectural critic of the nineteenth century, wrote to his father of his first encounter of Verona he could not help but to place it in his mouth. He writes of Verona: 'I should like to draw all St. Mark's and all this Verona stone by stone, to eat it all up into my mind, touch by touch.'[9]

This is perhaps a strange thing to write to one's father—but Ruskin was not one to strictly abide Victorian standards. It is not the taste of Verona that Ruskin wants (and wants to share with his father) but rather its touch. He wants to feel it—a total feeling, a feeling only fulfilled in consumption—of all buildings and of all stones. It's a nonsense of course, but at the same time it's a very real desire—to hold all—to encompass. To touch all and all at once. Deleuze deals with a notion much like 'touching all at once' in terms of the notion of interiority and what he and Guattari were later to describe as 'appropriation proper'.

There is a certain confusion in the utilisation of the term 'appropriate', deriving, perhaps, from a conjugation of the verb 'appropriate' and the adjective 'appropriate'. This linguistic slip promotes considerations of appropriation of efficacy or any number of other value judgments that tie the notion to the jurisprudence of meanings and motives. To appropriate is, first and foremost, the event of consuming that which was otherwise belonged or integral to another body (discipline, field of knowledge, subject). This desire to appropriate, to swallow, is the desire of interiority and configures the body in terms of machines of capture. Those machines that presuppose possession: to take, to have, to hold, to control, to be the one who possesses the entirety of the map, every stone of the city, the grand and godlike overview.

For Deleuze it was necessary to differentiate appropriation proper from 'encastement'.[10] 'Encastement' in Deleuze (and Guattari's) schema refers to the method by which a state keeps a warring body or class in relative isolation from state intentions. In referring to 'appropriation proper', Deleuze and Guattari are suggesting the full disciplining of one body by or into another (the disciplining of the warrior under state codes for example).[11]

The key difference between appropriation and encastement is that appropriation implies transformation. When Ruskin places Verona in his mouth, he conceives of a transformed Ruskin and not merely a consumed Verona. This characteristic of appropriation is identified in Charles Comte's early definition of appropriation in *Treatise on Property* (*Traité de la propriété* (1834)) where he suggests, 'by this action (of connecting) he appropriates (things) to himself. He transforms them into a part of himself, in a way that one could not detach them from him without destroying him'.[12] The present chapter will suggest that the work of Reinhardt and Loke involves multi-modal appropriations, appropriations that are not only about interiorising or swallowing but also about that which may in the end transform, burst, or destroy. Reinhardt and Loke's sensate machines are not mild; they 're-pattern the brain'. The appropriations of one body into another also serve as a means to exteriorise, to dance. The relation between bodies activates desires to exteriorise that which is more often than not considered to be part of a body's interior and its depths. Ruskin puts Verona in his mouth not only to swallow it but also to spit it back (in this case toward his father).

In *The Logic of Sense*, Deleuze suggests, in a manner that resonates with Comte's definition of appropriation, that '[t]o eat and to be eaten—this is the operational model of bodies, the type of their mixture in depth, their action and passion, and the way in which they coexist with each other'.[13] He aligns eating (as a way of being) with bodies (on a Stoic register); with causes ('all bodies are causes'[14]); with interiority ('the interiority of destiny as a connection between causes'[15]); with manifestation ('"manifesters" like the special particles I, you, tomorrow, always, elsewhere, etc.'[16]); with the state of affairs (the 'spatio-temporal realization' of the event[17]); with the alimentary system (alimentary orality, 'absorption and even excretion');[18] and, on a spatial register, with depth ('Orality, mouth and breast are initially bottomless depths').[19]

To consider then what constitutes and connects to speaking in *The Logic of Sense* is now far too simple an act. When speaking is constructed in series against eating, it is too easy (or too dialectical) to invert the list of that which aligned with eating and to attribute the inversion to speaking. However, a simple act of inversion would seem to be the case, even when the outcome is to articulate the difference between the series and not to invest in the dialectic itself. Thus, speaking is aligned: with the incorporeal (on the Stoic register); with effect ('effects are not bodies, but properly speaking, "incorporeal" entities');[20] with exteriority ('incorporeal effects are inseparable from a form of exteriority');[21] with denotation ('in the form of "it is that", "it is not that" ');[22] with the event ('the event belongs essentially to language');[23] with semiological orality (the vocal with respect to the oral); and, on a spatial register, with the surface ('by following the border, by skirting the surface, one passes from bodies to the incorporeal').[24]

One swallows and appropriates to be at one with the machine like a character in a Kafka novel where, according to Deleuze and Guattari: '[t]he mechanic is part of the machine, not only as a mechanic but also when he ceases to be one'[25]—the mechanic and the machine at once—appropriation proper—alimentary absorption. However, Deleuze and Guattari assert that though the characters of Kafka's novels are considered part of the machine of the novels themselves in terms of their work, leisure, loves and idiosyncrasies, they are also *mechanics* of the novels in their ability to assert a mode of agency; to transform as they are transformed—not only to be absorbed but to absorb—to speak and to spit. This is also the character of Reinhardt and Loke's sensate machines.

For, across the long and invertible lists created in *The Logic of Sense* and headed by the verbs 'eating' and 'speaking' Deleuze inscribes a line of mixes, of cuts and bruises—of quasi causes between causes and effects;[26] of borders and strips (and, importantly, surfaces themselves) between interiority and exteriority;[27] of a 'neutral' sense between manifestation and denotation;[28] of disequilibriums between the state of affairs and the event;[29] of esoteric words and exoteric objects between alimentary and semiological orality;[30] and, on the spatial register, of membranes—between depths and surfaces.[31] In addition to these cuts and bruises, we get notions that precede their concept of nomadism: 'places without occupants' and 'occupants without places'.[32]

Appropriation (swallowing) need not be aligned exclusively to the cavity of the mouth. Juhani Pallasmaa, the Finnish architectural theorist, most famous for his phenomenological text *The Eyes of the Skin* (1996) makes a parallel skirting of the surface and then a plunge for the bodies depths as Deleuze. In *The Eyes of the Skin*, Pallasmaa pays particular attention to the ear. He carves space from the ear— or rather, he allows the ear to carve its own space:

> Anyone who has become entranced by the sound of water drops in the darkness of a ruin can attest to the extraordinary capacity of the ear to carve a volume into the void of darkness. The space traced by the ear becomes a cavity sculpted in the interior of the mind.[33]

The ear is particularly inoffensive. It is the organ that gives us the aural and allows us to orient ourselves—and, by doing so, frames both sounds and bodies to spatialise and coordinate thoughts. The ear is defined by its internalisations—by that which it takes in, by that which it swallows. The image that comes to mind is of a dancer moving only to music and not to other bodies. When one dances like this it is often with the eyes shut. The ear is a beautiful tool for Pallasmaa—the phenomenological impulse to internalise and to contest a real external is played out in the ear. The ear is safe because it produces nothing and cannot bite back, spit or vomit. As long as one neglects the production of wax, one is completely secure in that the ear will assert absorption and embodiment without conceding the exteriority of the corporeal. The ear will swallow and swallow alone.

There are of course far more dangerous and exciting organs for those concerned with determinations of space and the exteriorisations of the body than the mouth or the ear. In *The Logic of Sense* though the mouth comes to be that cavity of negotiation between the dualities of speaking and eating and surface and depth, Deleuze acknowledges the schizophrenic disjunction of orality: 'the duality of things/words, consumptions/ expressions, or consumable objects/expressible propositions. This duality between *to eat* and *to speak* may be even more violently expressed in the duality between *to pay/to eat* and *to shit/to speak*'.[34] If we must have a duality (if only to bruise and to cut) then let us at least have an intense one.

And what is more intense than to shit/to speak and the mouth-anus? The anus implication of the mouth-anus is of course far more problematic and productive for architectural theorists—phenomenologists in particular. The anus is the site where not only do we have depth, but we have an excess of effect: of output. From the anus we have production— and what it produces is not of the surface. For Deleuze 'excrements are always organs and morsels' and therefore related to depth.[35] In a generation where architects (and artists) are concerned with the auto-poetic, with self-generating forms, with exteriority and with endogenous forces, one might have imagined the anus to be the body-zone par excellence of architecture—the image of production to which all truly autonomous architecture aspires. But it is not. It is not for lack of diagnosing the problems of interiority in architecture. In a text of 1992 titled *Monsters of Architecture*, the architectural theorist Marco Frascari introduced the problems of the body in terms of depth and surface through a consideration of the grotesque and the monstrous:

> The logic of the grotesque image ignores the smooth and impenetrable surface of the neoclassical bodies, and magnifies only excrescences and orifices, which lead into the bodies' depths. The outward and inward details are merged. Moreover, the grotesque body swallows and is swallowed by the world. This takes place in the openings and the boundaries, and the beginning and end are closely linked and interwoven.[36]

Though, for Frascari, 'the grotesque body swallows and is swallowed by the world', this monster of architecture is substantially under-fed and over-encaste.[37] The consuming phenomenology of the text *Monsters of Architecture* exposes the monster, however fertile in the textual morsel, to the contemporary preoccupation of affirming the health of architecture's interior. Frascari's monster may have had orifices but its excrescences were silent.[38]

The problem of the mouth-anus is, with little doubt, the problem of architectural production and its interiority. But, what is this interior of architecture? When the much famed architect/architectural theorist Peter Eisenman engaged with Noam Chomsky's linguistics in the early 1970s and in a paper 'From Object to Relationship II: Giuseppe Terragni—Casa Giuliani Frigerio' (1971), he was concerned with the issue of *deep structure*—though only skirted the surface.[39] Eisenman thought that, to give a framework for the proper articulation or 'speaking' of architecture, it was crucial to use a structure and he turned to linguistics for an apposite process.[40] He makes the imperative point, nonetheless, that by definition the 'objects' of language (words) have communicable meanings, and their structure in sentences generally follow exacting regulations. That is, they operate as categorisations rather than determinations. Architecture's 'objects' (material tectonics), however, are without 'agreed upon or intrinsic meaning'.[41]

Eisenman drew upon Chomsky's system as outlined in *Syntactic Structures* (1957). The system comprises three mechanisms: first, the *phrase-structure mechanism*; second, the *transformational mechanism*; and third, the *morphophonemic mechanism*. Each of these mechanisms define a set of laws that work upon a specified 'input' to produce a specific 'output'. Through a reading of Chomskian linguistics, Eisenman investigated the need to detach syntax from semantics in architecture. To do this he initially identified a configuration in architecture that corresponded to that of linguistics and denoted two structures: *surface* (percept) and *deep* (abstract). The descriptors thus bare little resemblance to Deleuze's usage.[42] Eisenman defined surface and deep structures in the syntax of architecture.[43] For him, 'deep structure is implicit only; it is not expressed but only represented in the mind'.[44] This approach 'worked' temporarily for Eisenman as a means to negotiate the speaking of architecture perhaps only because the architecture to which it was applied preceded the approach itself.

In a 1998 paper, 'It's Out There … The Formal Limits of the American Avant-Garde', the architectural critic and theorist Michael Speaks suggests that, '[t]his metaphysic of architecture is arranged around what Eisenman today calls an interiority, or by any other name an epistemology or ideology'.[45] It is the teleology of architectural production and the way in which the issue of 'form' tends to present itself as an interior limit-condition for architecture that is Speaks' concern. That is, when architecture (what it is and does) is the pre-ordained outcome of a process, then the process of architectural production is teleological, intentional and aligns it with the more carefully controlled outputs—speaking as opposed to shitting.[46]

For Reinhardt and Loke, interiority in the form of *deep structure* implies, as it might for Deleuze, a categorisation, striation, control, a zone of private property and a compartmentalisation of subjectivity. Reinhardt and Loke's work deals implicitly with Deleuzean notions of 'machine', 'sensation', and the 'becoming'. However, the interiority of architecture as a task often means that notions of pure exteriority, of exteriorised depth, of exteriorised corporeality (as opposed to the type of exteriority implied in speaking) remain notional and with little relation to the architectural project. It is at this point that Reinhardt and Loke's work threatens to burst a disciplinary boundary of architecture.

Reinhardt and Loke's most interesting spatial engagements are those referred to as the 'sensate' and the 'monstrous'. Sensate machines such as *Black Spring* are an attempt to exteriorise depth from architecture, to extract forces from its inside. According to Reinhardt and Loke, sensate machines are deployed against 'a representational view of the world or mere information processing'.[47] In *Black Spring*, Reinhardt and Loke evoke a type of space, a spatial determination, that superficially sounds very much like the product of Deleuze's mouth-anus. In *Anti-Oedipus*, Deleuze and Guattari suggest that:

> [P]artial objects are the molecular functions of the unconscious … If it is true that every partial object emits a flow, it is also the case that this flow is associated with another partial object and defines the other's potential field of presence, which is itself multiple … The synthesis of connection of the partial objects is indirect, since one of the partial objects, in each point of its presence within the field, always breaks the flow that another object emits or produces relatively, itself ready to emit a flow that other partial objects will break.[48]

The partial object is not a controlled and articulated product, but rather a self-generated sentience. For Reinhardt and Loke the sensate machine generates a self-identity that is 'embodied, embedded, enacted, extended, affected'.[49] I'll quote Reinhardt and Loke's own description that promotes the potential alignment between sensate machines and autonomised partial objects;[50]

> *Black Spring* is generated as a complex, open arrangement of different component groups (topography, flowers, audience) that are partially programmed, and partially interactive. Swarm behaviour is here extended to a maximum of component parameter that invests in complexity, temporality, and asynchronicity. Beyond maintaining a geometrical position, characteristics, or assignment, the flowers move erratically to an invisible code. A dancer bends and twists, establishing a second layer of audience engagement through subversive behaviour. The prosthetic/parasitic flowers refuse support function and demonstrate a life and will of their own; the dancer registers as a hybrid between topography and device; thus affecting practice and progression of the choreography. The procession of the installation is terminated through dissipation, the exhaustion point of animated system members.

In a paper written by Reinhardt and Loke and expansively titled 'GOLD (Monstrous Topographies)—Exploring Bodies in Complex Spatiality: Trespassing, Invading, Forging Body(ies)' the inclination is to consider corporeality as arising 'when the process(es) of understanding, perceiving and enacting the self, body, identity and spatiality become negotiable.'[51] However, Reinhardt and Loke's description of the spatiality of the sensate machines is predominantly geometric. It is a description based on centres, surfaces, and other spatial categorisations (not determinations). Reinhardt and Loke apply to these categorisations a notion of a 'diagram of forces' for their sensate machines. They appropriate the deformable geometry of the early twentieth-century Scottish biologist D'Arcy Wentworth Thompson as the best expression of the flow of forces upon form. It is this fluid geometry that possesses the ability to register particular localised transformations of a force on (and from) a 'sensate machine'. Reinhardt and Loke's *Black Spring* and its forces is an attempt to reconcile form with a dynamic bodily context. *Black Spring* is a signifier of an external context characterised as the supple expression of forces.

Reinhardt and Loke's understanding of force is anexact[52], unencumbered by immutable conditions of formal relation and pragmatic in its effect. The 'forces' of Thompson and Reinhardt and Loke do not compromise the homogeneity of the body due to: for Thompson; 'the intermolecular force of cohesion';[53] and, for Reinhardt and Loke, 'body constellations and identities: complex corporealities' that operate as flexible gravities assigned to each element. It is this supposed *need* to hold together, to control space, that is at once a very basic architectural drive, and also disturbing when thought through in terms of the potential of Deleuze's mouth-anus notion. The sensate machine still involves particular controls and internal gravities—reinforcing interiors and interiority—making wholes of part-objects—simultaneously guaranteeing what is produced whilst creating the 'new' as a pole and not an expectation. There is spitting here, but it is well aimed.

Our consideration of the entrancing productions of Reinhardt and Loke perhaps end at a place not unlike that of Deleuze, where the difference is not really between speaking and eating but rather between organisation and the body-without-organs; between interiorising depths and the exteriority of depths. In *The Logic of Sense*, Deleuze writes:

> The entire system of introjection and projection is a communication of bodies in, and through, depth … In this system of mouth-anus or aliment-excrement, bodies burst and cause other bodies to burst *in* a universal cesspool.[54]

Architects have always occupied the mouth. To eat and speak is to order the disordered and chaotic—to turn matter into output, work and energy directed toward objectives. There is in architecture a desire to formulate spatial categorisations from spatial determinations. Here the idea of a determination might best be thought of as a type of endogenous force that resists control. The teleology of architectural production—locates its outputs in the self-consciousness of the mouth[55] with the will to control what is expressed. This is probably why, in *A Thousand Plateaus*, Deleuze and Guattari explicitly align architecture and cooking as state endeavours.[56] The investments of Reinhardt and Loke are about giving over such control over exteriorisations—to produce but not to be in complete control of that production much like the productions of Deleuze's mouth-anus. Architectures of the anus, spatial determinations of the mouth-anus, would remain a production of partial objects, of the exteriorised corporeal, and be ever incomplete—to be fit to burst.

ENDNOTES

1   Gilles Deleuze, *The Logic of Sense*, edited by Constantin V. Boundas (New York: Columbia University Press, 1990). Translation of *Logique du sens* (Paris: Les Editions de Minuit, 1969), by Mark Lester and Charles Stivale.

2   Gilles Deleuze and Félix Guattari, *A Thousand Plateaus: Capitalism and Schizophrenia* (Minneapolis: University of Minnesota Press, 1987), 69. Translation of *Mille plateaux*, volume 2 of *Capitalisme et schizophrénie* (Paris: Les Editions de Minuit, 1980) by Brian Massumi. On Deleuze's concept of 'concepts' and their relation to metaphor refer to: Patton, Paul, 'Strange Proximity: Deleuze et Derrida dans les parages du concept', *The Oxford Literary Review* (18, 1996): 117–33.

3   Gilles Deleuze and Félix Guattari, *What is Philosophy?* (New York, Chichester: Columbia University Press, 1994). Translation of *Qu'est-ce que la philosophie?* (Paris: Les Editions de Minuit, 1991) by Hugh Tomlinson and Graham Burchell. Reference is to Jean-Jacques Lecercle, 'The Pedagogy of Philosophy', *Radical Philosophy* 75 (January–February 1996): 44.

4   Deleuze approaches the politics of spatial determination from a number of different relative directions in the final section of *A Thousand Plateaus*, '1440: The Smooth and the Striated': in terms of technological, musical, maritime, mathematical, architectural and aesthetic models. Deleuze and Guattari, *A Thousand Plateaus*, 474–500.

5   Dagmar Reinhardt and Lian Loke, 'Not What We Think: Sensate Machines for Rewiring'. In Proceedings of the 9th ACM Conference on Creativity & Cognition. Sydney, Australia: ACM, 2013, 135–144.

6   In addition to Reinhardt and Loke, the exhibition catalogue acknowledges the contributions of James Ye Won, Lee Stickells, Jonathan Fernandes, Marjo Niemela, Elmar Trefz, Alex Jung. Thanks was also expressed to Rod Watt, Rachel Couper, Ellen Rosengren-Fowler, Lisa Fathalla, Melinda Wimborne, Ingrid Pohl, Oni Oost, Kate Dunn, Chris Law, Ian Blampied, and the MARC4003 students.

7   Dagmar Reinhardt and Lian Loke, 'GOLD (Monstrous Topographies)—Exploring Bodies in Complex Spatiality: Trespassing, Invading, Forging Body(ies)', *International Journal of Interior Architecture + Spatial Design*, Spring issue II (2013).

8   Deleuze, *The Logic of Sense*, 223.

9   Edward Tyas Cook, *The Life and Works of John Ruskin* (London: Methuen, 1983) I, 263

10   Deleuze and Guattari, *A Thousand Plateaus*, 419.

11   Deleuze and Guattari, *A Thousand Plateaus*, 419.

12   Charles Comte, *Traité de la propriété*, 2 vols. (Paris: Chamerot, Ducollet, 1834), 51.

13   Deleuze, *The Logic of Sense*, 23. Reference is particularly to Alice's coronation dinner.

14   Deleuze, *The Logic of Sense*, 4.

15   Deleuze, *The Logic of Sense*, 6.

16   Deleuze, *The Logic of Sense*, 13.

17   Deleuze, *The Logic of Sense*, 53.

18   Deleuze, *The Logic of Sense*, 198.

19   Deleuze, *The Logic of Sense*, 187.

20   Deleuze, *The Logic of Sense*, 4.

21   Deleuze, *The Logic of Sense*, 169.

22   Deleuze, *The Logic of Sense*, 12.

23   Deleuze, *The Logic of Sense*, 22.

24   Deleuze, *The Logic of Sense*, 10.

25   Gilles Deleuze and Félix Guattari, *Kafka: Toward a Minor Literature* (Minnesota: University of Minnesota Press, 1986), 81. Translated from *Kafka: pour une littérature mineure* (Paris: Minuit, 1975), by Dana Polan.

26   Deleuze, *The Logic of Sense*, 8–9.

27   Deleuze, *The Logic of Sense*, 10–11.

28   Deleuze, *The Logic of Sense*, 14, 19.

29   Deleuze, *The Logic of Sense*, 65.

30   Deleuze, *The Logic of Sense*, 51.

31   Deleuze, *The Logic of Sense*, 103–4.

32   Deleuze, *The Logic of Sense*, 41.

33   Juhani Pallasmaa, *The Eyes of the Skin: Architecture and the Senses* (Chichester: John Wiley and Sons, 2005), 50.

34   Deleuze, *The Logic of Sense*, 85.

35   Deleuze, *The Logic of Sense*, 189.

36   Marco Frascari, *Monsters of Architecture: Anthropomorphism in Architectural Theory* (Maryland: Rowman & Littlefield Publishers Inc., 1991), 32.

37   Frascari, *Monsters of Architecture*, 32.

38   That Frascari predisposes his concept of architectural monster to the further encastement of a phenomenological framework is illustrated by an enthusiastic review of the text by Perez-Gomez; Perez-Gomez, Alberto, 'Monsters of Architecture', book review, *Journal of Architectural Education* (46/1, September 1992): 60.

39   Peter Eisenman, 'From Object to Relationship II: Giuseppe Terragni—Casa Giuliani Frigerio', *Perspecta* (13–14, 1971): 38–61. Chomsky postulated a syntactic foundation of language and he called this 'deep structure'.

40   Rosalind E. Krauss, 'Death of a Hermeneutic Phantom: Materialization of the Sign in the Work of Peter Eisenman', in Peter Eisenman's *Houses of Cards* (New York: Oxford University Press, 1987), 166–84.

41   Eisenman, 'From Object to Relationship II', 39.

42   This point is explored in greater depth in Megan Yakeley, Deconstruction for Everyone: The Myth Exploded (University of East London, MSc thesis, 1992).

43   Eisenman, 'From Object to Relationship II', 39.

44   Eisenman, 'From Object to Relationship II', 39. It must be noted that later, however, in a lecture to the Yale School of Architecture in 1978 Eisenman declared the attempt to find syntactic meaning in the Casa Giuliani Frigerio to be 'hopeless'; cited in Yakeley, *Deconstruction for Everyone*.

45   Speaks, 'It's Out There', 30.

46   Daniel W. Smith, suggests that '[o]ne cannot have pre-existing norms or criteria for the new; otherwise it would not be new, but already foreseen': 'Deleuze and the Liberal Tradition: Normativity, Freedom and Judgement', *Economy and Society* 32(2), (May 2003): 308.

47   Dagmar Reinhardt and Lian Loke, 'Not What We Think: Sensate Machines for Rewiring'. In Proceedings of the 9th ACM Conference on Creativity & Cognition. Sydney, Australia: ACM, 2013, 135–144.

48   Gilles Deleuze and Félix Guattari, *Anti-Oedipus* (Minneapolis: University of Minnesota Press, 1983), 324–25. Translated from *L'Anti-Oedipe* (Paris: Les Editions de Minuit, 1972), volume 1 of Capitalisme et schizophrénie, by Robert Hurley, Mark Seem and Helen R. Lane.

49   Reinhardt and Loke, 'Not What We Think: Sensate Machines for Rewiring', 137.

50   Reinhardt and Loke, 'Not What We Think: Sensate Machines for Rewiring', 142–143.

51   Reinhardt and Loke, 'GOLD'.

52   Introduced by Husserl, 'anexact' refers to geometric forms that have measurable properties at a local level and yet are irreducible and therefore unrepeatable. Edmund Husserl, Cartesian meditations: An introduction to phenomenology (Kluwer Academic, 1960).

53   Thompson, *On Growth and Form*, 17.

54   Deleuze, *The Logic of Sense*, 187.

55   Deleuze, *The Logic of Sense*, 79–80.

56   Deleuze and Guattari suggest 'that architecture and cooking have an apparent affinity with the State, whereas music and drugs have differential traits'; Deleuze and Guattari, *A Thousand Plateaus*, 402.

Figure 5.1: *Black Spring/biome*, 2012, Tin Sheds Gallery, The University of Sydney.
Photo: Baki Kocaballi.

ENTANGLED: COMPLEX BODIES AND SENSATE MACHINES

DAGMAR REINHARDT
LIAN LOKE

The body is a machine. The machine is a body, a body-machine-architecture where conceptual inversions destabilise accepted notions of body, of identity, of architecture, and of space. Software code, computer programs, and computational scripts are the new languages of generating form, of creating new relationships, of dissolving boundaries. They shift static architecture into choreographed spatial environments. In this new design landscape where computation and materiality collide, the boundaries between bodies, machines, identities and environment are called into question.

Situated at the intersection of architecture, choreography and human–computer interaction, our work explores a generative concept of the body for understanding and inventing new relationships between complex spatial structures, performance, sensation and computation. Whereas previously the digital and the physical realms could be approached separately, now computational design processes give birth to new physical forms, and equally allow embedded computation to inhabit material objects through programmed material behaviour and performance.[1] As a consequence, our bodies are reconstituted as the environments we inhabit, and the objects with which we interact exhibit new properties and behaviours of an animated, kinetic nature. The body is not simply the literal, physical human body, an instrument to occupy a preferential plane.[2] The body extends itself to include virtual, animated and fabricated bodies (architectural elements, computation and electronics) in a complex body–machine symbiosis.

As part of our design research, we explore how a number of different bodies can be made to interact with each other so as to form and invent new body constellations and identities—complex corporealities—within novel spatial environments for creative engagement and expression. The creation of spatial practices imbued with expressive behaviours offers new modes of co-existence between humans and machines where creative engagement and sensual meaning-making are prioritised.

We present here a framework that drives our research, an under current shared in the creative works of three spatial installations—*The Black Project: Black Spring*[3] and *Black Shroud*,[4] and *GOLD (Monstrous Geographies)*.[5] These installations act as laboratories developed over a design, construction and experiential period within an exhibition

Figure 5.2: *Black Spring/biome*, 2012, Tin Sheds Gallery, The University of Sydney. Photo: Baki Kocaballi.

Figure 5.3: *Black Spring/biome*, 2012, Tin Sheds Gallery, The University of Sydney. Photo: Baki Kocaballi.

Figure 5.4: *Black Spring/biome*, 2012, Tin Sheds Gallery, The University of Sydney. Photo: Baki Kocaballi.

context and through exposure to an uninitiated audience. This essay opens a conversational field in which we operate—by laying bare a matrix of relationships that oscillates between a number of conceptual centres: the idea of complex corporealities; the idea of interdisciplinary design processes as re-invention; the idea of anexactitude and excess; and the idea of sensate machines.

COMPLEX CORPOREALITIES

We understand bodies as complex, malleable, contingent, multiple entities. Body complexity arises out of encounters; of bodies meeting with material entities, encounters choreographed through code. Bodies are informed and receive corporeality by an action or event that evolves within a spatial setting[6] through temporal and durational characters of performances. We seek non-normative environments for re-learning who we are and what we can become; a temporary 'slackening of the intentional threads',[7] which embed us tightly in the world; spaces of potential, infiniteness, and performative ambiguity in contrast to an architectural economy of user profiles, in contrast to spaces subservient to sensory capacities that safeguard body identities. As bodies enter a reciprocal relationship when space is internalised, the identity of a person is continuously formed by exposure, restrictions, actions and sensations, which are tightly bound to Deleuze's figure and fact.[8] An individual person approaches spaces, objects, and other bodies from identity, feelings, experience and memories. The approach evokes a strategy for dissolution in engagement— by sensation, affect—through shifting material presences and combinatory identities. Complex corporealities arise by shifting sensation and experience into unknown territories.

To charge the capacity of a material to change in the process, dissolving and recombining audience or agent (human response) in performance, we deploy software programming behaviour and performativity of matter. Through insertion of these animated, responsive and spatialised bodies, a complex system of hybrids arises in which bodies are crossing over, are programmed or linked together, creating a balance, eruption and sensation that renders bodies of the individual participants (and that includes the installation's bodies as much as those of the audience and actors) as 'prostheticised', contextualised, extended, affected. In this manner, the ownership, faculty, permanence and identity of the individual body is challenged in multiple acts of sharing, dissolving, trespassing, reconstructing engagements, sensations and experience. Bodies may then be constituted by performance; a performance charged by software. Multiple identity emerge through sums of parts[9] so as to feed cognition, and capacity for learning is fast-forwarded by arousing curiosity.[10]

Our grouped, animate, prostheticised, experiential bodies are coded through serial computation processes: designed, scripted, fabricated, constructed and programmed—as composites of physical form, materials and computation. They seamlessly and repeatedly cross boundaries between analogue and digital, actual and virtual realms. Developed in the computational realm of 3D modelling software, they reconnect to that virtual topography through (software, behavioural) code that allows these animate bodies to contribute kinetic, autonomous and responsive behaviour. This choreography thus sets the temporal presence of performers and audience into relation with the designed, manufactured and programmed body entities so as to stimulate, invigorate and translate an electronic dataflow into a sensate response and change behaviour—for the bodies we habituate. In the next sections, we describe the enactment of our concepts of corporealities into an interdisciplinary design process, and into an applied practice realm.

ANEXACTITUDE AND EXCESS [INCLUDING OVERABUNDANCE, DECADENCE]

We embrace excess, overabundance, decadence, delay, and morphological shifts. While design regularly argues for a purity of conceptual departure, we are interested in the layering, the traces, the assemblages that connect ideas and their formal expressions to an existing context. Excess thus relates to a richness of layers as much as a multiple repetition of a formal language. Excess and assemblages of bodies can produce moments of possibilities; that is, act as points of departure for experience of architectural environments.

Contemporary architecture uses the tools of computational generative design to produce striking new spatial forms. Generative design derives highly complex, non-typological morphologies of an architectural object that is the result of a formation process,[11] by which different internal and external forces (e.g., contextual change, growth, pressure, erosion) can interact on the same object, leading to transformations and evolutions of form. Open in natural evolution, in architectural practice this 'systemic delay' of design variations is necessarily confined prior to construction,[12] then allowing no latency, ambiguity or capacity for change. This effectively contributes to an 'ethic of statics',[13] by which the architectural body is privileged over the human body occupying the spatial envelope. We apply a purposed *anexactitude* that crosses through material expressions and open behaviour: anexactitude is by no way an approximation; on the contrary, it is the exact passage of that which is underway.[14]

Morphological variations in contemporary architecture can be interpreted as overabundance in a design process, yet remain restricted because they are not executed. In contrast, a strategy of ambiguity, or shape anexactitude, re-opens morphological processes to user impact and engagement. The base mechanism of formulating a body, and relationships between different bodies of the same kind or other species, can be enabled through rethinking latency in material: a materiality of latency for spatial complexity.[15] This strategy allows different descriptions of material formations or behaviour, different techniques for structural or organisational specifications, and can act as initiative for experiential effects and sensations.

A second strategy for exceeding a single instantiation of form is pursued through programming materiality with dynamic behaviours by forecasting morphological shifts through systems open to adopt contextual forces.[16] These computational programs have the capacity to be continually updated and re-programmed, thus extending morphological variations through an excess of programmed, kinetic behaviours. A choreographic approach enables thinking across and through heterogeneous materials, forms and expressions, to consider the temporal organisation of relations in terms of flows, transitions, transformations, interactions and autonomous behaviours.

Choreographic strategies produce traces that give threshold conditions for specific 'states' in the system: critical 'states' of the topography and body states. A performance is just one possible enactment, suggesting an array of possibilities for engagement. The installations we develop are concerned with the latent potential released through human interaction, open forms where opportunities for creative engagement and multiple levels of meaning-making are an integral aspect of how the design is conceived.

THE BLACK PROJECT AND GOLD [MONSTROUS GEOGRAPHIES]

Complex corporealities, processes of re-invention, anexactitude and excess, and sensate machines are interpreted in our installations *The Black Project* and *GOLD*. *Black Spring* tested a topography of multiple differentiated material systems programmed through diverse behaviour of one actor.[17] *Black Shroud* revised the previous *Black Spring* through more homogeneous groups in a more choreographed organisation.[18] *GOLD (Monstrous Geographies)* applied *The Black Project* design collaboration and optimised choreographic toolset to complex prototype groups and multiple actors.[19]

BLACK SPRING

Black Spring (figure 5.4) functions as an interactive environment responsive to audience presence, with a separate mode for choreographed performance. It is composed of a digitally fabricated topography constructed of timber and perspex, connected to an array of suspended polypropylene 'flowers', the movements of which can be controlled through programmable electronics. Its topography bridges between the virtual and actual by acting as a relational composition used to simulate material form generations; and as a programmable landscape that establishes control over a sensate environment. Two sets of data intersect here. Firstly, contour lines on the fabricated manifold (laser-cut topography) organise material response, by which the segmented planes allow the surface to respond in degree to adjustments in spacing (height/width of voids) in space (set by the installation field). Secondly, an imprinted system of perforations acts as a reference point for the 'virtual' topography that organises the flower modules, and enables these modular elements to respond to program signal, in height changes. The flowers are a diverse component group assembled of minor variations of petals, partially equipped with reflective stems (perspex) and light emitting sources (LED lights). These are connected to a set of servo-motors mounted above the structure, which pull tension cables that are threaded through the vertical planes of the installation. The relationship between the tension cable and motor allows for an actuation (in adjustment in height, by degree) that takes into account the variable individual expression of floral bodies, their weight, directionality, dimension and rotational capacity.

The bridge between the physical installation and its virtual counterpart (the programmed behaviour and response) is enabled through continual actuation and interaction through motion and movement. In the first phase of development, motion in the flowers is deployed as a continual internal re-calibration of a movement that corresponds to the structural deformation through gravity. It is an autonomous kinetic behaviour, with no input from the environment. The potential for engaging curiosity is limited, but can be further extended through adding interactivity. In the second phase, the behaviour of the installation is stimulated by a choreographed body and its states of posture and dissolution. The space was divided into four distinct spatial zones used to trigger different machine responses based on how we programmed the use of the input data. The sensing zones could be resized and were used in conjunction with a choreographic toolset. A Kinect motion sensor acts here as the input device by skeleton tracking, and its data is used to vary the kinetic behaviour of the flowers that respond in swarm-like behaviours (with a solitary flower presiding over a segment of the landscape, apart from the swarm).

BLACK SHROUD

Black Shroud reworks a continuation and revision of *Black Spring* with decreased project parameters. Whereas *Black Spring* used a complex topography designed between the programmed behaviour and a digitally manufactured topography, this installation instead uses a clean canvas of a grid (2m x 2m) that is connected to an array of identical polypropylene 'flowers', whose movement is electronically programmed, and responsive through motion sensors. This work equally uses the transfers between virtual and actual environments, yet more significance is given to the temporal relationships in the network. A homogenous component group is assembled of multiplied flowers from identical petals, and hung in identical height relative to a virtual reference plane.

Where in *Black Spring* a multiple complex network is initiated, *Black Shroud* seeks to establish a quiet, meditative environment that is unfolding as a choreographed movement between the dancer, its memorised traces and the spontaneous enactment by the audience (and potential clashes between). Emergent patterns of interaction and behaviour generated by the audience may deviate from the original choreography. This prototype links the embodied movement of a dancer in a delineated space via embedded movement through programming to an enacted responsive behaviour of the flowers. A Kinect motion sensor acts here also as the input device, where data are used to link the kinetic behaviour of the flowers to the memorised choreography of the virtual landscape. Multiple pathways of motion (dancer, flowers, audience) are thus set in correspondence with each other, traced, memorised and actuated. Each time new movement is mapped onto the experiential space of the spectator, cognitive behaviour is affected while passing through the choreographed space.

Both prototypes of *The Black Project* employ components engineered to work together seamlessly to support a performance setting, in which the notion of 'becoming part' with the (digitally manufactured, programmed and animated) environment through a 'sensate machining' acts as focus. In both installations, behaviour is explored through an interdisciplinary, iterative and collaborative exchange between the disciplines of computational architectural design, human-computer interaction and choreography.

GOLD [MONSTROUS GEOGRAPHIES]

GOLD (figure 5.5) expands the corporeality and spatiality of bodies from actual levels towards a conceptual level shared between play narrative, spatial diagram and interactive conditions. Marcel Duchamp's *The Bride Stripped Bare by Her Bachelors, Even (The Large Glass)*, 1915–23, serves as base for the three-dimensional translation into a choreographed space of the performance: a context of a four-dimensional spatiality, in which sets of bodies are linked through diagrammatic layers in multiple exposures of identities and personalities. *GOLD* transfers bodies and relationships in the *Large Glass* to three distinct body groups and their interactions with a self-organising multiple audience body that refreshes with each performance: the 'Brides'; the scissors and crowns (a number of prosthetic mechanisms); and the Fleur-de-Lis (swarm components) that interact with the bodies of actors and audience.

The performative capacities of human and animate (artificial) bodies in *GOLD* are played out through a number of different choreographic sets of movement that can act as a framework with which to organise the relationships between bodies, and generate temporal assemblies of meaning and interpretation. The 'Bride' is expressed as three golden clouds (inflatables produced from emergency blankets), which are highly responsive to existing contextual wind forces, and subject to direct tactile impact of the actors (and which are territorial in the sense that they delineate the topography of the performance). While the clouds remain passive-aggressive, both the prosthetic body group and the swarm bodies are prompted to organise in behavioural movement. The swarm component body (laser-cut modules of mirrored perspex) is a series of morphological variations of a single base type of a Fleur-de-Lis. Coupled with a servo motor actuation to correspond with the musical score, it maintains distinct behavioural capacity that alternates from autonomous (pre-coded moving sequences) to kinetically responsive formations that accompany the musical score, similar to the embedded choreography in *The Black Project*.

The Black Project and *GOLD* are examples of our designs for sensate machines: machine as architecture, anarchitecture of machines, of desire, curiosity. They collapse traditional boundaries between humans, machines and space, not by human–machine interfaces, but as human–machine–architecture configurations—an evolving pattern of relationships and interaction. Our machines are complex and ambiguous, with a hidden agenda, recruiting human agency, stimulative in contact and exposure.[20] Non-trivial machines, non-purposed, non-functionary, but stimulative and behavioural; that is, opening or inviting to an experience, a sensation. They exploit threshold conditions through direct haptic contact (hand, handle, touch) in which body and machine become one organisational unit/organism. No longer machinic in concept, they echo delirious apparitions,[21] cut to the experiential core of materiality, spatiality and corporeality.

We approach machines as kinetic entities that connect to human behaviour, and machines that are diagrams, generators and stimulators of architectural discourse, instrumental to actualisation, prompting new realities.[22] These sensate machines are producing bodies, producing relationships, connecting behaviours: body and behaviour becoming available to the machine (rather than servicing a user). As an architectural practice, they negotiate previously set contour lines that interpolate between space and body, thus resetting the facts of the occupational figures that explore these new realms. They bring forward a delay—a moment to re-assess, re-constitute, re-live—relieve.

The Black Project acknowledges singular body entities. Yet, in *GOLD*, a corporate hybrid body understanding is the main driver, with a capacity for serving prosthetic functions to other animate or human entities: a body group contains different species of prosthetic scissor mechanisms (based on a Chuck Hoberman Sphere) that are assembled as different entities (the 'Crowns' and 'Spiders'), and which signify disabling, enhancing, or replacing aspects of relationships. Complex corporeality is driven here by anexactitude/material latency for composite, choreographable bodies specifically through the way in which these bodies can be constituted in morphological variations. These mechanistic/prosthetic bodies are generated through a generic scissor mechanism, whereby six major components constitute a body; the 'Spider', whose major characteristic is the ability to un/fold (in different positions due to its mechanical nature). This morphology is then extended to formulate a more complex body/ies: the 'Crown'. Inbuilt in the Spider's mechanical composite body is a potential to adapt to spatial context due to its *mathematical* coding. More complex formations combine the former 'Spider's' three major 'limbs' in a hexagonal pattern in a code for self-forming systems. These spiders receive their spatiality, their bodily identity, through different stages of expansion and contraction through a coded elastic response to formative (choreographed) requests of their bodies in space. The crowns and spiders with their kinetically designed expansion/contraction and programmed behaviour refer to a literal (physical) expandability. Yet, notions of elasticity may also be deployed as choreographed changes of matter in time through morphological shifts. Similar to their

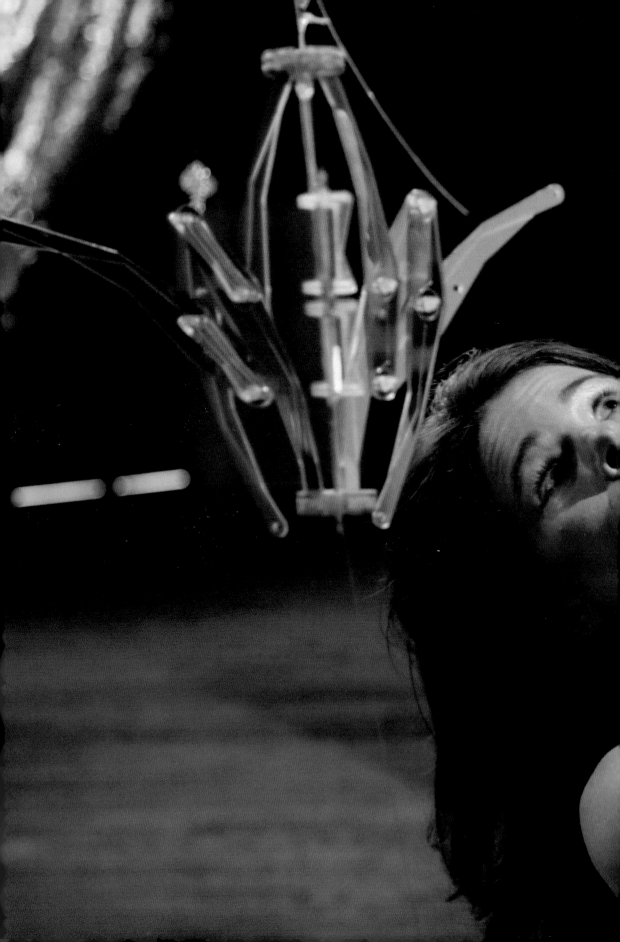

biological precedents, these animalistic entities are the results of forces and resistances other than generic types; they define the duration of a form through alternatives. Gradient conditions of the performative context shape the individual form of which each is a result of duration; that is, a process version informed by specific tuning of choreographic criteria. Differences can thus be traced through the deformation through body parts, which correspond to individual (actor, audience) bodies. These deviations seem to simulate a 'species' adaptation to changing contexts, a change of material form effected by the movement of matter.

In *GOLD (Monstrous Topographies)*, a phenomenal elasticity may be considered to facilitate a shift in coding and alterations: of the individual identity and subjectivity; of the body; or of the act of spatialisation. When character properties and programmatic protocols are shifted, then body or space may be reconfigured, recombined and reinterpreted. Individual perception relates to the self, to what is consciously perceived by experience. This opens a discourse that allows departures from predetermined expectancies and contingencies, because a complex corporeality interpolates between the interiority and exteriority of the self, between form and content, between the appearance and identity of bodies and objects, species and spaces.

RE-INVENTION IN INTERDISCIPLINARY DESIGN PROCESSES

We design research: as a speculative assertion (design is a verb).[23] Design research includes creative acts that develop dialogue, doing, multi-direction, and processes of investigation that privilege deviations, sideways and alternative results. For us, design is research, specifically in a context of spatial and corporeal complexity. Explorations span body, space, code and materiality in excess of architectural practice's privileging of form, function, economics; thinking through designing with concepts, diagrams, methods, process, in many media through which design is pursued and produced. Our design process combines iterative cycles of phases of analysis (observation) and synthesis (making), reflection, adaptation, and connaissance (enjoying the moment). We directly apply conceptual approaches through an understanding of concepts as 'shining points of multiplicity', which remain approximate and multiple, and change with the notions involved. Not singular but multilayered, outside a constructed systematicity.[24]

As a primary agent for our creative work, we aim to unveil a multiplicity of conceptual and semantic layers within a potential work; layers folded within layers, narratives and plateaus that bridge between architecture, engineering, interaction design, programming and choreography—between literal and abstract, virtual and physical, analog and digital.[25] We play with association, multiple layers of meaning, quotations of signage and science. Taking a familiar sign and distorting it introduces layers of ambiguity so as to open access to meaning, again through non-representational means. The means include latency that eschews programmed interactions through wildcards, cycles of intuitive audience engagement, taking the individual in open-path, open-source space and time.

Figure 5.5 *GOLD (Monstrous Geographies)/ILoveToddSampson*, 2013, Pier 2/3 Walsh Bay Sydney. Photo by Nathaniel Fay/oneskidigital. Performer: Gabrielle Quin.

We approach design research from a dialogue between different disciplinary fields through intersected and recombined processes of language, invention, design methodologies, conceptual approaches, and multiple layers. Our thinking frequently involves combining concepts or elements: a collaborative exchange as a creative act in which separate areas, lenses, and thinking heads are regularly interlinked, exchanged, exposed, contaminated and cross-bred. We learn through processes of collaboration and confrontation, developing a language of exchange between disciplinary fields as a critical engagement that fosters movement, rather than containment or control.[26] Colloquial terms applied in one discipline change in meaning (problem of semantics): code and programming in architecture do not mean the same in interaction design. Architecture can refer to the architecture of a system or process as much as it refers to a static environment. We draw out layers of connotation, of meaning, through our own discourse of intervention.

ACKNOWLEDGEMENTS

This ongoing research between the Computational Design Unit and the Design Lab on complex corporealities has been supported by the Faculty of Architecture, Design and Planning, the University of Sydney, through the Zelda Stedman Bequest. Installations were fabricated with support of the Digital Fabrication Lab of ATSC at the University of Sydney. We also acknowledge our research collaborator, Alexander Jung (reinhardtjung architects, www.reinhardtjung.de), and our installation specific team members: *Black Spring* (Dagmar Reinhardt, Lian Loke, Marjo Niemela, Alex Jung, Elmar Trefz, Jonathan Fernandes, James Yee Won Lee); *Black Shroud* (Dagmar Reinhardt, Lian Loke, Alex Jung, Chris Law, Ingrid Pohl, Jodie McNeilly); *GOLD* (Dagmar Reinhardt, Lian Loke, Alexander Jung and reinhardtjung architects, Eduardo Barata and UFO Sydney, Elmar Trefz and semnon, James Yee Won Lee, plus 2012 Master of Digital Architecture Research students at the University of Sydney, Nicole Larkin, Kristy Whiting, Thomas Murray, Dominik Broadhurst, Zainab Tinwala, Tom Novakovic, Alina Minassian).

ENDNOTES

1   Branko Kolarevic and Ali Malkawi, ed., *Performative Architecture: Beyond Instrumentality* (New York: Spon Press, 2005).

2   Maurice Merleau-Ponty, *Phenomenology of Perception* (London: Routledge, 1962).

3   Black Spring, Biome, Tin Sheds Gallery, Sydney, August 2012 (Collaborative Team: Dagmar Reinhardt, Lian Loke, Marjo Niemela, Alex Jung, Elmar Trefz, Jonathan Fernandes, James Yee Won Lee, Chris Law).

4   Black Shroud, Organized Cacophony, Pop Up Gallery, The Rocks, November 2012 (Collaborative Team: Dagmar Reinhardt, Lian Loke, Alex Jung, Chris Law, Ingrid Pohl).

5   GOLD (Monstruous Geographies), ILTS, TheLivingRoomTheatre, Pier 2/3, Walsh Bay, Sydney, March 2013 (Collaborative Team: Dagmar Reinhardt, Lian Loke, Alex Jung, Eduardo Barata, Elmar Trefz, James Yee Won Lee).

6   cor·po·re·al *adj.*1. Of, relating to, or characteristic of the body, synonym bodily. 2. Of a material nature; tangible. www.thefreedictionary.com/Corporeality.

7   Merleau-Ponty, *Phenomenology.*

8   Gilles Deleuze, *Francis Bacon: The Logic of Sensation*, trans. Daniel W. Smith (London: Continuum, 2003).

9   Steven Johnson, *Emergence—The Connected Lives of Ants, Brains, Cities and Software* (London: Penguin, 2001).

10   John Protevi, 'Deleuze and Wexler: Thinking Brain, Body, and Affect in Social Context'. In *Cognitive Architecture*, ed. Deborah Hauptmann et al. (Rotterdam: 010 Publishers, 2010), 168–84.

11 Achim Menges and Michael Hensel, *Versatility and Vicissitude* (London: Wiley, AD, 2008).

12 Ali Rahim, 'Contemporary Processes in Architecture', *Architectural Design* 70 (2000).

13 Gregg Lynn, *Animate Form* (Princeton: Princeton Architectural Press, 1998).

14 Gilles Deleuze and Félix Guattari, *A Thousand Plateaus: Capitalism and Schizophrenia* (Minneapolis: University of Minnesota Press, 1987), 21.

15 Elizabeth A. Grosz, 'Future of Space'. In *Architecture from the Outside: Essays on Virtual and Real Space* Elizabeth Grosz (ed.) (Cambridge: MIT Press, 2001), 109.

16 D' Arcy Wentworth Thompson, 'On the Theory of Transformations, or the Comparison of Related Forms'. In *On Growth and Form* (Cambridge: Cambridge University Press, 1917).

17 Lian Loke and Dagmar Reinhardt, 'First steps in body-machine choreography' (workshop proceedings 'The Body In Design', Melbourne, 26–30 November 2012).

18 Lian Loke and Dagmar Reinhardt, 'Not What We Think—Sensate Machines for Rewiring' (paper presented at the ACM Creativity and Cognition Conference, Sydney, 17–20 June 2013).

19 Lian Loke and Dagmar Reinhardt, 'GOLD (Monstrous Topographies)—Exploring Bodies in Complex Spatiality: Trespassing, Invading, Forging Body(ies)', *International Journal of Interior Architecture + Spatial Design*, Spring issue II (2013).

20 As described by Heinz von Foerster, 1967. Cited in Steven Gage, 'The Wonder of Trivial Machines'. In Robert Sheil, *Protoarchitecture: Analogue and Digital Hybrids*. (Vol. 78. John Wiley & Sons, 2008), 4.

21 Alphonso Lingis, *Foreign Bodies* (New York: Routledge, 1994), 21.

22 Deleuze and Guattari refer to the diagram as an abstract machine that 'does not function or represent, not even something real, but it builds the real to come, a new type of reality.' Deleuze and Guattari, *A Thousand Plateaus*, 157.

23 Peter Downton, *Design Research* (Melbourne: RMIT University Press, 2003), 92.

24 Robert Ausch, Randal Doane and Laura Perez in an interview with Elizabeth Grosz at City University of New York Graduate School and University Center, Kansas State University, accessed 30 August, 2013, www.k-state.edu/english/symposium/2004/speakers/grosz.html.

25 Grosz, 'Deleuze and Space', lecture held at the University of Sydney, lecture notes received as mail correspondence of author (10 September 2003).

26 Chris L. Smith, 'The Problem of Writing Architecture: Knots, Restraint and Pleasure', *Journal for Cultural Research*, 10 (2), (2006), 185–97.

6

INVENTION AT THE NETWORK SCALE

[NETWORK CONFIGURATIONS]

INVENTIONS ARE NETWORKS:
FOSTERING THE LIMINAL PLAY OF IDEAS

SEAN LOWRY

EXPANDING SONIC SPACE: AN ANTIPODEAN
APPROACH TO TELEMATIC MUSIC

IVAN ZAVADA

INVENTIONS ARE NETWORKS:
FOSTERING THE LIMINAL PLAY OF IDEAS

SEAN LOWRY

It is not simply that networks enable invention. Inventions are in themselves networks. Inventions do not arrive in the world autonomously and fully realised. By contrast, novel devices, methods, concepts, compositions and processes are invariably formed as contextually dependent adaptations, improvements or hybrid manifestations of pre-existing webs of devices, methods, concepts, compositions and processes. Moreover, the meaningful recognition of any invention's value is also contingent upon pre-existing networks of contextual relations. Consequently, this double state of dependency upon networks implies that all inventiveness is at least indirectly collaborative. This collaborative spirit of inventiveness roundly encapsulates the human ability to take existing objects and ideas and combine or repurpose them in new ways. Given that this process can occur in both geographically distant real time and across historically asynchronous moments, digital networks are clearly capable of radically accelerating the mating of ideas. Our urge to participate and collaborate within and across networks is being fundamentally reinvented within an ethos of the digital. This chapter will discuss various ways in which global networks are liminally expanding and re-choreographing the irresolvable yet productive tensions between individual and collective interests that underpin cultural change, and therefore, by extension, drive invention.

Humans are social creatures, and our inventiveness is bred via collaboration. This collaboration can take many forms. It can occur via real time correspondence and feedback, via trans-historical or trans-cultural referencing and appropriation, or via wholly unexpected collisions between ideas previously distant in time and space. Within the now broadly recalibrated and poststructuralist infused rethinking of the notion of creativity that has emerged in the aftermath of the so-called 'death of the author', the myth of the individual standalone genius has been both heavily discredited and substituted for a more contextually contingent interpretation of inventiveness. Accordingly, emphasis has also shifted from an immovable and essentialist sense of the supposedly inherent value of ideas toward an appreciation of the contextual significance of an idea's reception and interpretation. After all, it is in the context of an invention's reception (often located elsewhere in time and space to that of its conception) that value is finally attributed. It is networks that provide the vehicle to connect these events. To this end, an absence of linkages can become a larger problem than an absence of ability. Without a network to nourish it, an invention will die or lie dormant. For US-based Hungarian psychology professor Mihaly Csikszentmihalyi, 'creativity is not simply a function of how many gifted individuals there are', but rather 'how accessible the various symbolic systems are'.[1] Or as US popular science author and media theorist Steven Johnson put it in his 2010 TED talk, 'chance favors the connected mind'.[2] Here, the point that Johnson is really emphasising is that, given an appropriate environment, ideas brought together can potentially become significantly larger than the sum of their parts.

New ideas have long emerged within multidisciplinary spaces. Just as Enlightenment thinking coincided within the rise of coffee houses[3] that brought people from different professional backgrounds together, and eighteenth century societies of learned gentlemen met to discuss topics as disparate as poetry, fossils and politics in a single social space, environments in which groups of curious people might come together in a context not necessarily defined by background, geography or disciplinary and vocational demarcation remains fundamental to the generation of new ideas. Invention at a network level is dependent upon connecting groups, clusters, precincts, cities or historical epochs such that people with different skills and expertise might think collaboratively. From Athens through the Renaissance to modernity, cosmopolitan clusters have given rise to innovation. Gradually, technological advances such as telegraphic communication would herald a more global sense of cosmopolitanism. Meanwhile, as new modes of rail, sea, road and air transportation arrived, and increasingly complex trading networks evolved, information about networks would soon become as valuable as physical commodities. Digital networks have simply expanded this phenomenon at a massively complex global scale. In hyperrealising the dynamic tensions that have long underpinned the contested domains of human interaction, the internet offers both a relatively deterritorialised space for free and open communication and a political instrument for both nation-states and multinational corporations.

The sharing of ideas is clearly a key ingredient underpinning inventiveness. No individual knows how to produce a computer, a space station, or a national constitution entirely on his or her own. Given that the informational components that make up a new invention often exist long before the inventor, inventions might be regarded simply as significant variations within inherited informational networks. It therefore stands to reason that the more prescriptively confined an inherited informational network is within a disciplinary silo or cultural context, the more limited its potential scope for influence and lateral dissemination will be. This is where the lateral connectivity of digital social networks coupled with the cross-disciplinary potential of open source digital databases can help to open up fields that are otherwise starved by narrow filters in terms of the recognition of novelty. Naturally, the collective likelihood of inventiveness is exacerbated by an increased interconnectedness of individual minds (each in and of itself a product of collectively generated experience). Unfortunately, given that secretive research environments still incentivise inventiveness, the private ownership of many of the constituent parts for potential inventions invariably limits the size of the gene pool (to employ the tempting analogy of natural selection). Open networks emphasise the value of sharing ideas rather than protecting them within patents or copyrights. The Open Invention Network (OIN), for example, is a company that acquires patents and then licenses them royalty free to entities that, in turn, agree not to assert their own patents. Fortunately, open networks that enable individuals to crowd source up-to-the-minute information are appearing in many spheres of human activity. Significantly, this rhizomatous climate is also both producing leaderless political movements (such as the Occupy Movement) and helping to transform broadcast culture and synchronous communication into opportunities for people to share their thoughts on any subject and develop their views through asynchronous discussion. Here, inventiveness might range from historically unprecedented transformation to the more modest context of an individual's life. In this sense, social networks are potentially promoting the production of singularities in a world that is otherwise increasingly homogenised (certain problems with this utopian claim are discussed later in this chapter).

Historically, a key problem that limits the development of new ideas is the incongruity of the historical or cultural context into which they are born. Ideas that are born without a supporting cultural infrastructure typically miss out on the transformative potential of innovation. Consequently, what a particular society values in inventiveness will directly influence the kinds of innovations that it produces. It is also difficult for any society to anticipate what future societies might value. Given that many inventions eventually serve a purpose never envisioned by their inventor(s), it is therefore manifestly sensible to maintain access to as many incomplete, unfunded, untenable and unrealised inventions as possible such that future (and geographically or culturally distant) generations might exercise their judgment. In short, ideas that are potentially well suited for mating can be simply out of synch with one another. To this end, databases, discussion boards, digital archives and social networks can help to facilitate asynchronous exchange across time and space. An incomplete archival thread on a discussion board, for example, might be picked up months or years later and then brought to fruition with the injection of new information. Significantly, as much as invention often unfolds within stimulated social and professional contexts, time lapsed connectivity can invite quieter moments of reflection to join the discussion. Alarmingly, the ephemeral and present focused nature of digital communication is unsympathetic to this potential.

To be recognised as novel, a creative idea must be accepted, in the words of US psychologist Morris Stein in 1953, 'as tenable or useful or satisfying by a group in some point in time'.[4] Consequently, the criteria for assessment must naturally predate an invention's reception. As a direct extension of their work, inventors must often play a role in actually establishing that criteria for assessment. French artist Marcel Duchamp, for example, made sure that he was right in the middle of the action when his early twentieth century invention of the 'readymade' became significant for a new generation of post–World War II artists. Then, as the velocity of idea transmission accelerated exponentially via new communicative technologies through the second half of the twentieth century, the networked reach of Duchamp's invention proliferated in virtually every node of the art world. Ultimately, networks not only increase the chance of an invention being united with an appropriate context of reception, they can also enable inventions to be united within unexpected contexts. Wherever access to information is lost, the potential for later generations to develop ideas that an intermediary generation had discarded is also potentially lost.

The process of invention is often an exploratory process leading to uncertain or unknown outcomes. Without accepting failure as a possibility, scientific research could not take place. Israeli chemist Daniel Shechtman, for example, freely admits that his research that culminated in the discovery of quasiperiodic materials was successful precisely because he was given the freedom to 'prepare materials which were strictly useless'.[5] Cultural invention is no different. Some failures remain spectacularly rooted as historical moments and others fall from history. Yet remaining mindful of the value of the history of failure both insures against repetition and inspires recalibrated reassessment. Risk-taking inventiveness is often a precursor to 'successful' and 'useful' innovations. Given that the distance between the events of invention and innovation can be enormous, and given that the girth of human activity across the present is starting to dwarf the cumulative weight of history (arguably fostering a worrying historical amnesia), the intelligent use of databases and networked connections is becoming increasingly vital to the challenge of insuring that these events remain linkable.

Despite the reality that many inventions are popularly linked to a specific generative context, they are often also both tacitly indebted and potentially transposable to other disciplines. In any case, disciplines are only subcategories within the collective culture that produces and validates them, and are therefore open to liminal play and cross-fertilisation. Although the inventor, innovator and producer can be discrete or equivalent agents, it is in the strategic exploitation of an appropriate context of reception that the most historical noise is typically generated. After all, it is at the point of social validation that an invention is either recognised as knowledge or adopted as a form of behaviour that is actually passed on to others. Although distinctions between the contexts of discovery and social validation are blurred, the value of an invention is certainly significantly indebted to its validating context. In this sense, Duchamp, for example, is indebted to the later generation of artists that actually recognised value in his invention of the 'readymade'. The important thing is that ideas find an appropriate network within which to fruitfully unfold and be meaningfully shared. It is for this reason that invention and innovation are concepts that remain inextricably intertwined—for in order for inventions to be evaluated, they must be made manifest. Given that that the process of making an invention manifest requires at least some degree of innovation, the task of distinguishing inventiveness from innovation is clearly problematic. It is also difficult to distinguish 'successful implementation' from peer validation. To make such a distinction, for example, within any acknowledgement of Kazimir Malevich's enormous legacy to abstract painting or John Cage's legacy to avant-garde music would be absurd. Here, the utilitarian conception of 'successful implementation' is simply not part of the *raison d'être* of advanced art. In this context, ongoing peer validation is a more meaningful measure.

Today, the accelerated reach of global networks certainly makes inventive collaboration possible in a meaningful way well beyond the physical limits of geographical specificity. Countries once geographically isolated, such as Australia, can now communicate cheaply and instantaneously with the fastest growing parts of the world. Yet, at the same time, partisan political and disciplinary climates often promote polarities. Although the tyranny of distance is evaporated, intellectual laziness and historical amnesia can still limit inventiveness. Although, for example, Australia's economy has largely sidestepped the global financial crisis thanks to its minerals wealth, education and innovation continues to slide in a culture that is not yet incentivised to value invention. Meanwhile, Australia's domestic markets and cultural networks are so small that the additional challenge of internationalisation must be met. Clearly, to effectively participate globally, digital connectivity and meaningful access to information is fundamental.

Information is ultimately only as effective as its relational positioning in a network of connections. This relational positioning is of course also in a constant state of flux. The open nature of the digital networks emphasises cultural forms that are incomplete, unresolved and open for constant addition, editing and remixing. Meanwhile, the internet has transformed the world into an enormous dematerialised library built within the metaphor of the cloud—albeit one that remains substantially limited by the editorial choices of its contributing individuals. In the words of legal researcher Lawrence Liang, 'the utopian ideal of the library … like its predecessors in Alexandria and Babel … is destined to be incomplete, haunted by what it necessarily leaves out and misses'.[6] The key point of difference that underpins networked inventiveness in the digital age is the searchable interconnectedness of networked databases containing useful information. Databases help to facilitate invention by linking information, ideas, people, and special interest groups together in ways that transcend anything that could ever be 'borne of mind' or made possible within the traditional geographical and historical limitations of human interaction. In the words of British computer scientist Tim Berners-Lee, best known as the inventor of the World Wide Web, the most important use of data is 'by somebody else for some reason that you've never imagined'.[7] By extension, for Berners-Lee, the ' fact that when you connect two pieces of data together that you get much more value means that as more data is produced … the value of each piece goes up'.[8]

Significantly, databased informational relationships are not limited by traditional linear narrative as a form of perception. Networked databases are enormously customisable in terms of drawing relationships between informational elements. Consequently, information can now be meaningfully sorted and organised within radically vertically and horizontally expanded relational structures and in accordance with as many thematic connections as there are individually circumscribed journeys through these ever-expanding options. The impact that this is having upon the sheer weight of human activity across our present is substantially transforming traditional notions of subjectivity and historical perception. Significantly, given that the database represents the world as groups of items as opposed to a representation in terms of 'a cause-and-effect trajectory of seemingly unordered items (events)',[9] the database has become a 'natural enemy' of narrative in contention for 'the same territory of human culture'.[10] But, given that databases only provide meaning via the subjective interpretation of information, software capable of organising and connecting subjective interpretations— i.e,. social, professional, and political networks—is clearly an essential part of the equation. Considering that we now more readily accept reality as 'constituted by a multiplicity of spatialized temporalities'[11] and history as moving at differing and overlapping velocities, individuals nonetheless experience subjectively distinct relational possibilities that are continuously recycled through the accumulation of collective culture across time and space.[12]

Under the dual influences of digital reproducibility, which allows information to be reformatted and disseminated effortlessly, and the exponential acceleration of exchange enabled by globalisation, individual subjectivity is responding to the dynamically collective power of digital networks in fascinating ways. Extending the algorithmic power of search engines and previously unimaginable correlations, human subjectivity continues to make its mark in terms of selecting, arranging, sorting, recontextualising, reformatting, interpreting and analysing data in novel ways. Relatively freed from remembering 'facts', which although contingent upon multiple interpretation and political bias are for the most part searchable (and in the interests of cultural, political and historical transparency should set off alarm bells wherever they are not), our role are arguably increasingly shifting toward an emphasis upon the importance of understanding relationships between facts, where and how to find information, acknowledging the potential for gaps and omissions, and perhaps above all an ability to transpose all of the above into new and emerging contexts. This is also arguably the challenge for twenty-first century education—to foster conceptually transposable forms of inventiveness that are potentially applicable to contexts not yet known or imagined.

Forever flattening distinctions between producing and consuming, and increasingly utilising customisable access to databased information, inventiveness is now often constituted in terms of constantly recalibrated 'versions' that bear no singular or 'finished' destination. In a network, cultural objects are not stilled but rather subjected to ever-changing contexts of use and consequently constantly redefined in their translations from place to place. At any rate, although the way in which we experience any creative work has always been contingent upon networks of experience, culture, religion, class, and politics etc., the way in which the open nature of digital networks emphasise cultural forms that are never resolved is exponentially hyperrealising this contingency. For the digitally networked inventor or cultural producer, a studio or production space is now likely to be a laptop in the kitchen or local coffee shop that is connected to a networked space shared by innumerable other 'studios', potentially shifting a physically solitary experience into a more explicitly collaborative environment. In addition, the site of production is typically the same as that of dissemination and reception. Given that digital networks can flatten former distinctions between producing and consuming, new virtual worlds of endlessly remixed versions are being inhabited by new generations of so-called prosumers (a term coined by US futurologist Alvin Toffler in 1980). It would appear, for example, that most fans of highly specialised 'micro-genres' of electronic music are those also DJ'ing or producing within the same subgenre.

The multiplicitous historical trajectory that is transforming formal disciplinary distinctions into contextually contingent interdisciplinary inconstancies now seems irreversible. Consequently, with specifically produced and pre-existing elements attributed relatively equivalent roles within the construction of meaning, definitions of inventiveness are also becoming increasingly contextually dependent. Herein lies a paradox—fractured interdisciplinarity is resulting in new hybrid forms of specialisation. Within the same capitalist culture that specialisation is still widely understood as a natural response to the way in which the social order favours the most efficient use of limited quantities of time and capital, much cultural production is simultaneously becoming multi-skilled, interdisciplinary and autonomous. Where popular musicians are likely to also be writers, recording artists, promoters and live performers, graphic designers are expected to be web designers, metrics analysts and entrepreneurs—all tasks that would have once more likely have been demarcated. In this sense, key points of difference are potentially less likely accorded to cultural producers possessing remarkable ability in one domain, but rather a creative understanding of their relations.

An unlikely outcome of this aforementioned paradox is that ever-smaller subcategories of specialisation are forming at previously unimaginable points of hybridity in the multiple margins of collapsed disciplinary centres. These newly minted specialised interdisciplinary configurations then invariably fragment further, resulting in smaller but globally connected networks. Consequently, the only individuals interested in consuming this very specific new knowledge are those working within these newly minted specialised interdisciplinary configurations. For the new partisan, the subtlest of permutations can once again contain specific interest, and, once again, those who attempt to translate the new micro-field into something widely communicable are treated with suspicion. Consequently, in an age of informational abundance, the construction of individually circumscribed prejudicial boundaries promotes (an allusion of) diversity. Within the context of contemporary art, for example, US art historian Rosalind Krauss has observed a shift that she has describes as a 'post-medium condition'; that is, something that actually 'heightens, rather than reduces, the importance of the concept of the media', and, moreover, re-presents the idea of the 'medium as an aggregative "network" or "complex" of media'.[13] German art critic Jörg Heiser has employed the term 'super-hybridity' to describe hybridised forms of cultural production that transcend 'the point where it's about a fixed set of cultural genealogies', and instead become ' a kind of computational aggregate of multiple influences and sources'.[14] Australian art critic Edward Colless has nominated a similar phenomenon using the term 'superabundance'.[15] At any rate, once a creative work enters a network of collective consideration it cannot remain stilled. Instead, it is subjected to myriad contextually nuanced incarnations, dislocations, fragmentations and degradations. The broader the network, the less likely it will remain quarantined inside disciplinary boxes.

Some creative challenges and problems require many different disciplines. Consequently, it is important for potential collaborators to find new ways to transcend disciplinary specialisation and narrowcast communication. Invention and innovation require a dynamic balance between individuals who possess depth in specific areas of expertise, and individuals who understand how things fit together and can work across disciplinary boundaries by analysing ways in which specific knowledge might relate to broader spheres of knowledge. Given that inventive thinking often involves combining concepts from previously unrelated realms of knowledge, recognising a new possibility, connection, or relationship in and of itself clearly constitutes a form of inventiveness. Inventiveness across disciplines is starting to challenge long held assumptions in universities that inventions are only based in science or engineering. According to President of Stanford University John L. Hennessy, for example, the 'arts can help us break out of traditional patterns of thinking', for 'discontinuous innovations' require 'more cross-disciplinary collaboration'.[16]

Amplified and reiterated in networks of mass communication, digitised information can rapidly become unanchored from its contexts of origin. It is becoming increasingly difficult to comprehend the complexity of the invisible networks that now link us together in such an abstracted way that formerly localised events can exacerbate to a global economic calamity at a previously unimaginable velocity. Clearly, the scale and complexity of digital networks eclipse anything that an individual can possibly comprehend. Consequently, instead of attempting to comprehend the overall contours of these networks, we are arguably better served in thinking about the kinds of movements that characterise these networks. It is here that the idea of transitivity becomes significant. Given that objects within a network are characterised by their circulation and translation into new contexts, transitivity is about recognising inventiveness as something that becomes apparent in flux. The expectation that an invention will inhabit a singular destination (within a patent, copyright or art museum, for example) can ossify potential for ongoing translation. US scholar and art critic David Joselit employs the term 'transitive painting' to describe artists who[17] 'actualiz[e] the *behavior of objects within networks*'—i.e., perpetual circulation and retranslation— by staging the 'passage' from 'painting-as-cultural artifact to the social networks surrounding it'.[18] As Joselit reminds us, many artists now recognise social relationships as aesthetic extensions of an artistic practice. Over the last two decades in particular, many artists and theorists have worked toward aesthetically framing spheres of social relations. French curator Nicolas Bourriaud, for example, has laid claim to the term 'relational art' to describe 'a set of practices which takes as their theoretical and practical point of departure the whole of human relations and their social context'.[19] This is not, however, an entirely new idea. From 1950s and 1960s French situationism through to the rhizomatic participatory art movements of the 1960s and 1970s, advanced art has long traded medium specificity and disciplinary categorisation for a critical reflection upon the networked nature of contextual relationships. Similarly, the meaning systems that underpin many fields of invention are now generally more reflexively dependent upon a critical relationship with the broader semiotic systems that structure society.

Just as artists take art outside of its a traditional contexts in order to exploit other means of dissemination, strategic planning in any field involves analysing interconnections—both in terms of how one part of an organisational unit fits another, and the way in which that organisational unit connects with other organisations, researchers, developers, suppliers and customers. In any case, the interaction of various parts of a society and way in which inventions are created and exploited within that society is inextricably interrelated. If one connection is cut, a whole system is affected. It is therefore important to try and understand how a system fits together by mapping and identifying the mechanisms needed for adjustment. Systems theory is an interdisciplinary study that aims to elucidate principles potentially applicable to all types of self-regulating feedback systems. Based on general systems theory, as developed by Austrian biologist Ludwig von Bertalanffy, systems theory is an acknowledgement that relationships between objects perform a significant function in addition to objects in and of themselves. A systems conception of inventiveness should therefore accommodate tensions between individual agency and the strictures of collective contexts and institutional structures. Given that cultural invention is cultivated in the productive yet ultimately irresolvable tension between individual and collective demands,[20] the exponential scale of global networks is certainly radically accelerating this phenomenon. The complex interplay between the spheres of an individual's habitus and their social context, coupled with the accumulated knowledge of collective endeavour both across and through time, is what makes invention both possible and meaningful. With so much symbolic capital dependent upon the value of peer recognition, the ability to maintain relations within any social or professional context is clearly critical for any invention's 'success'. It is certainly not insignificant that the term 'networking' itself is typically used to refer to the exchange of information or services between individuals, groups, or institutions, and the subsequent strategic cultivation of productive relationships.

Given that inventions are dependent upon networks for both conception and reception (a condition hyperrealised in late twentieth century mass communication and made ubiquitous in the digital age), it is interesting to note that conceptual artists such as Hans Haacke, Daniel Buren and Andrea Fraser have long worked toward inventing various ways of aesthetically problematising this phenomenon. Since the 1960s, a second (and initially less prominent) branch of conceptual art developed an institutional critique rooted, in part, in the idea of 'systems aesthetics'. Systems aesthetics, unlike the more production and process-oriented focus of much conceptual art, is explicitly concerned with circulation, organisation, feedback and broad interrelationships in economic, social and linguistic systems that stretch way beyond the circumscriptions of the art world. Here, the worlds of communication, data, and media spaces become an aesthetic medium. For Jack Burnham, one of the main forces behind the emergence of systems art in the 1960s, this was a new world in which transformation 'emanates, not from things, but from the way things are done'.[21] Similarly, inventiveness in its broadest sense can be modelled as multi-scaled and dynamically responsive at micro, meso and macro levels.

Another line of questioning that has variously appeared since the 1960s is concerned with relationships between natural systems and human networks. To what extent can ecology and technology be modelled as interrelated parts of complex systems? After key graduates of post-1960s counterculture (the generation that at least rhetorically sought a balance between technology and nature) moved on from tuning in and dropping out to create Silicon Valley, we face new and seemingly insurmountable challenges at the networked frontiers of eco-techno complexity. In the words of Joselit, Google Maps already 'turn up Star Trek–inspired images of interconnected solar systems that do little to enhance one's understanding of the traffic in information but do much to tie digital worlds to ancient traditions of stargazing'.[22] As we might analyse the past, the task of speculating about the future has proven consistently problematic. After all, to repurpose the words of German critic and philosopher Walter Benjamin, 'the angel of history' faces 'toward the past'.[23] Within complex interconnected systems, unexpected events can certainly bear enormous consequences. We have no idea which new technologies and markets will emerge or the gravity of the challenges and opportunities we will confront. We can, however, aim to develop forms of thinking potentially transposable toward an unknowable future. For Lebanese-American author Nassim Nicholas Taleb (an avowed enemy of vocations of prediction), we can prepare for 'unknown unknowns' by accepting that option-like experimentation outperforms directed research—despite the way in which we rationalise with hindsight.[24]

As traditional perceptions of historical time diminish, the girth of collective activity across our present moment is expanding toward new levels of informational obesity. For popular US media theorist Douglas Rushkoff, 'presentism' has become the ethos of a society always turned on, operating and updating live and in real time, but ignoring the more nuanced transformative potential of asynchronous digital communication.[25] Arguably, although the most transformative technologies of our time are the smartphone and mobile computing, it is neither the old world synchronicity of the telephone conversation nor the new world up-to-the-minute banality of social media that really epitomise their transformative potential. By contrast, it is in the access to both asynchronous networks of useful information and the potential for realtime networked collaboration that transformative potential really begins. In the case study that follows Sydney-based composer Ivan Zavada explores the way in which network music collaborations, now no longer limited by geography or language, are redefining relationships between performer and performance environment to produce new forms of musical expression. In virtually any sphere of human activity, new possibilities are emerging. Flattening activities as disparate as researching, conferencing, banking, producing, consuming and performing into a individually connectable portable space, the increasing ubiquity of mobile digital devices might even help to transform invention in developing economies (particularly by availing access to micro-financing). We can only hope that a democratisation of digital communication technologies, coupled with the enormous growth of potentially connectable data and individuals working from disparate fields, places and times, might unlock an unprecedented capacity for free, open and truly global invention. At any rate, the challenge of fostering the inventiveness that humanity needs to either divert or adapt to the unknown calamities that an exacerbated level of activity in itself will bring will require individuals and collectives capable of transcending entrenched cultural, political, professional and disciplinary boundaries.

ENDNOTES

1   Mihaly Csikszentmihalyi, 'Implications of a Systems Perspective for the Study of Creativity', in *Handbook of Creativity*, ed. Robert J. Sternberg (Cambridge: Cambridge University Press, 1999), 333.

2   Steven Johnson, 'Where Good Ideas Come From'. *TED*. September 2010. Accessed 25 April 2013. www.ted.com/talks/steven_johnson_where_good_ideas_come_from.html.

3   Johnson, 'Where Good Ideas Come From'.

4   Morris Stein, 'Creativity and Culture', *The Journal of Psychology* 36 (1953): 311.

5   Dan Shechtman, quoted in: 'The Scientist Who Answered His Critics—By Winning a Nobel!' ABC Radio National, *The Science Show,* broadcast 30 March 2013. Accessed 25 April 2013, www.abc.net.au/radionational/programs/scienceshow/the-scientist-who-answered-his-critics---by-winning-a-nobel21/4607452.

6   Lawrence Liang, 'Shadow Libraries', *e-flux*, 2012. Accessed 28 April 2013, www.e-flux.com/journal/shadow-libraries/.

7   Tim Berners-Lee. 'Keynote Speech at the Launch Of CSIRO's Digital Productivity and Services Flagship', 14 March, 2013. Accessed 20 April 2013. www.csiro.au/en/Organisation-Structure/Flagships/Digital-Productivity-and-Services-Flagship/TBL-keynote.aspx.

8   Berners-Lee. 'Keynote Speech'.

9   Lev Manovich, *The Database, Language of New Media* (Cambridge, Mass. MIT Press, 2001), 218 -43. Reprinted in ed. Zoy Kocur and Simon Leung, *Theory in Contemporary Art Since 1985* (Minnesota: Blackwell Publishing, 2005), 408.

10 Manovich, *The Database*, 413.

11 Jan Verwoert, 'Apropos Appropriation: Why Stealing Images Today Feels Different', *Art & Research: A Journal of Ideas*, Contexts and Methods, 1 (2007): Accessed 25 April 2013, www.artandresearch.org.uk/v1n2/verwoert.html.

12 Sean Lowry and Ilmar Taimre, unpublished working paper, 2010.

13 Rosalind Krauss, *A Voyage on the North Sea: Art in the Age of the Post-Medium Condition* (London: Thames & Hudson, 2000), 56.

14 Jörg Heiser, 'State of the Art', Editorial. *Frieze*, 133.

15 Edward Colless, 'Superabundance', *Broadsheet: Contemporary Art & Culture*, 33(4), (2004), 16–18.

16 John L. Hennessy, The Role of Creativity and the Arts in a 21st-Century Education: President Hennessy's annual address to the academic council. April 20 2006. *Stanford News*. Accessed 21 April 2013, news.stanford.edu/news/2006/april26/hentext-042606.html.

17 David Joselit, 'Painting Beside Itself', *October* 130 (2009): 128.

18 Joselit, 'Painting', 128.

19 Nicolas Bourriaud, *Relational Aesthetics* (Paris: Presses du reel, 2002), 113.

20 Eleanor Heartney, 'In Praise of Uncertainty', 1993. In *Critical Condition: American Culture at the Crossroads*, ed. Eleanor Heartney (Cambridge: Cambridge University Press, 1997), 7.

21 Jack Burnham, 'System Esthetics', 1968. In *Great Western Salt Works: Essays on the Meaning of Post-Formalist Art*, ed. Jeffrey C. Burnham (New York: George Braziller, 1974), 160.

22 Joselit, 'Painting Beside Itself', 128.

23 Walter Benjamin, 'On the Concept of History', 1940. Translation: D. Redmond, 2005. Original: *Gesammelten Schriften* I:2. Frankfurt am Main: Suhrkamp Verlag, 1974, accessed 25 April 2013, www.efn.org/~dredmond/Theses_on_History.html.

24 Nassim Taleb, *The Black Swan: The Impact of the Highly Improbable* (London: Penguin Books, 2007).

25 Douglas Rushkoff, *Present Shock: When Everything Happens Now* (New York: Current, Imprint of Penguin Group, 2013).

EXPANDING SONIC SPACE:
AN ANTIPODEAN APPROACH TO TELEMATIC MUSIC

IVAN ZAVADA

Telematic music, also known as network music performance or net music, involves the use of telecommunication technologies to facilitate the synchronous exchange of audiovisual information for multimedia experimentation, musical performance and creativity over high speed communication networks such as the internet. As network music events are no longer contingent on geographical location nor language as the only method of communication, new and unique forms of artistic practices exist, and can be shared almost instantaneously and interactively over long distances to convey human expression and foster new realms of creativity. One of the main principles behind network music collaborations consists of extending the boundaries of conventional performance and creative practice in a society highly dependent on communication and network technology. Such extended interactive environments create drastic changes in the dynamics for collaborative artistic processes around the world. This case study explores the mechanisms redefining the relationship between performer and performance environment, and explores how collaborative processes generate new forms of musical expression. Network music performance creates a global opportunity to preserve, integrate, or modify cultural identifiers in an environment facilitating the exchange of cultural items, such as music and other forms of artistic expression.

Emerging technologies play an important role in shifting conventional music praxis towards multi-dimensional and multi-layered systems relying on flexible creative and conceptual mechanisms. Some of these mechanisms involve redefining the relationship between performer and performance environment and exploring the correspondences between different mediums of expression, while considering and exploring the juxtaposition and synthesis of regional aesthetics in the creation of new works.[1] The introduction of a communication network layer creates a flexible setting for improvisation and interactivity allowing musicians to expand their artistic and creative boundaries. In this new performance context, musical material often emerges from a non-theoretical framework, where musical form and structure is revealed through the creative process itself rather than pre-existing or conventional composition rules (i.e., musical score).

Music has always been a sharing experience, both on stage between performers and between performers and the audience. The communicative aspect of music takes a new dimension through the integration of networks and the internet, which has become a new creative resource in music and has made a drastic impact on the compositional experience where new parameters are to be considered in the drafting of creative activity.[2] Often composition is seen as an individual activity; the introduction of new communication methods immediately expands the creative potential for the composer and changes the role of the composer to more of a coordinator of musical events over time and space (geographical location).[3] This may also involve the coordination of several real-time musical events or pre-composed musical pieces performed simultaneously in different places, or even in proximity, relying on network technology.

The issue of synchronisation could be dealt with directly by considering delays between various locations and the time it requires for encoded sound to be transmitted across vast networks.[4] In the musical sense, synchronisation is bound by technological and traditional aspects of musical performance cues. However, it is part of the current network performance model to remain as flexible as possible due to latency issues. Logistical and technological aspects include the distance between performance locations, bandwidth of the transmission system, network protocol, signal strength and reliability of the network connection. Traditional aspects rely on the musician's ability to perform in remotely located spaces. This new setting therefore engenders non-conventional cues for performance, relying solely on sound for example, or removing visual or gestural cues to coordinate musical events.

Taking the above aspects into account in the creation of a new work would require a very precise calculation of musical event onsets and extraordinary performance accuracy across networks and would be barely achievable even with the fastest network connections. Since data transmission is limited by various factors such as bandwidth and latency of the system, even in an ideal high capacity data transmission scenario, transmission rates are dependant on light passing through fibre optic cables over long distances.

As an example, the distance between Sydney and Beijing is almost 9000 kilometres; it would take at least 0.03 seconds for light to travel that distance and even more for packets of data to be transmitted through that distance. Interestingly, this corresponds to 30 milliseconds, which is approximately the threshold at which humans perceive two different sounds and can distinguish echoes.[5] Distance can therefore be a new factor limiting perfect musical synchronisation. In a local acoustic environment, 30 milliseconds corresponds to a sound wave travelling approximately ten metres from a sound source to a listener's position. Performing via network between Sydney and Beijing in two different halls is therefore analogous to performing in the acoustic space only ten metres apart. This is in the ideal world and, from recent tests performed between the two cities, latency often varied between 50ms and 250ms. In a real-time performance situation, one must also consider that the encoded sound must travel back and forth to allow the collaborating performer to coordinate and synchronise musical gestures. It is therefore imperative to take latency variability into account when creating a new performance and consider it as a new parameter in the elaboration of a work.[6] Network latency can be circumvented through the invention of new musical paradigms to create or simulate a sense of rhythm rather than trying to perfectly synchronise musical onsets or incorporate syncopated rhythmic onsets/attacks. Other sound processing techniques may include various effects affecting timbre.

The first complex network music performance that occurred at the Sydney Conservatorium of Music was a performance involving an erhu player located in Calgary, Canada, and the concert took place at the Central Conservatory of Music of China in Beijing during the 'Musicacoustica Festival' in October 2011. It was the first time a group of students from Australia participated directly in a creative musical performance over Internet2 (IPV6 protocol). The event was collaboration between three institutions and was coordinated by Kenneth Fields, Canada Research Chair in Telematic Arts—Syneme.[7] The concept was to create a performance system where a traditional erhu player would perform in Calgary; the instrument's sound was captured by a microphone during the live performance and sent through the network to the computer music studios at the Sydney Conservatorium of Music. The sound was processed and transformed to create sonic effects in a computer lab and was sent back almost simultaneously to support the musicians' improvisation; to complete the loop, the sound was mixed and sent to the location of the concert in Beijing. The audience in the Beijing concert hall had a seamless listening experience and was not aware of the long distances travelled by the encoded sound stream from continent to continent. There would have been no perceivable time discrepancies in the music or combined sound streams due to the nature of the effects used on the erhu. Time lag depended mainly on broadband network speed, digital signal processing techniques on location, efficiency of the equipment used for signal processing and the proficiency of the hardware/software operators in the computer labs. Technical aspects such as network delay and synchronisation were therefore circumvented through the choice of effects, mainly affecting timbre (granulation, ring modulation, distortion, reverb) and therefore not time dependent conceptually.[8] Time-based effects such as delays and echoes were avoided due to the complexity of the network and the need for clarity in the musical material. The resulting textural material was used to support the live sound of the erhu, and provided a directional feedback in anticipation of forthcoming musical material to be improvised by the performer.[9]

A series of preliminary experiments illustrated that musical style and technological context both play an important role in determining aesthetic outcomes of a network music performance.[10] As such, live sound transformations and manipulations are not mandatory for a network music performance, but it would be rather disappointing not to integrate sound processing techniques in the signal chain. Improvisational proficiency also plays an important role in the musical flow. The musician was asked to choose and perform a reference piece representing her cultural background, or of socio-cultural significance. The chosen work was entitled *Mo Li Hua*, literally 'Jasmine Flower', and originated in the Jiangsu province on the east coast of China during the eighteenth century; the song is very popular and has been performed at major events and is therefore engrained in the Chinese culture. The performer is considered to be of advanced level and has been playing the erhu for 14 years. The piece was first played without any sound processing, mixing or other sound transformations, to focus on musical tradition and authentic sound of the instrument. A second version of the same piece was performed, but this time accompanied by signal processing techniques such as granulation (a form of time-stretching of an audio file or live sound input). Finally, the musician was asked to improvise, without taking into account the previous interpretations and try to be guided by the sounds accompanying the performance, both from the instrument and the sound effects generated by the computer—this was the more experimental version that would determine if the performer could break away from traditional performance aspects or, on the contrary, if new elements of creativity would arise as a result of the disconnect from the more rigid conceptual setting.

The flexibility of the musician to explore various playing techniques and detach themselves from previously learned musical habits or precepts is of the essence.[11] Musical precepts refer to stylistic embellishments naturally arising from previously learned performance mechanisms and reflexes derived from conventional or traditional music repertoire. One of the main challenges of a network music performance environment is to encourage spontaneity and interactivity in the performance, with the technology being as transparent as possible in the interaction process to ensure uninterrupted musical flow. This experiment revealed that spontaneity and interactivity in the performance as well as reciprocity in the dialogue between the performers resulted in unexpectedly unique musical outcomes and offered new conduits for artistic development, provided that adequate technologies are available for high quality transmission of audiovisual information.

These initial explorations with erhu led to further experimentation with multichannel sound and multiple layering of musical textures and sound effects. The next logical step was to include a visual component normally in the form of a score or musical instructions to be performed at the other end of the performance network. The use of conventional scoring techniques defeated the purpose of using a non-conventional performance system such as a network. Therefore, the new challenge was to transmit visual information that could relate musical information in a non-conventional way to generate performance events based on structured abstract information. Similar to the sounds of a dial-up modem or a fax machine, the crackling and hissing noises generated from the modem seem random and unstructured, yet the resultant information becomes intelligible if decoded and formatted within another system. An abstract visual score is somewhat analogous to these aforementioned sounds. The resultant imagery may seem unstructured and random at first, but once decoded into another realm such as music it can evoke very specific musical gestures and structures. The visual score encodes and at the same time suggests musical motifs and textures that can be adopted in various contexts, instrumentations and configurations. Complex imagery translates into multi-parametric musical information. Colour, shape, brightness, contrast, movement, direction, layers, symbols, space, speed, etc. are all variables that can be interpreted musically as pitch,

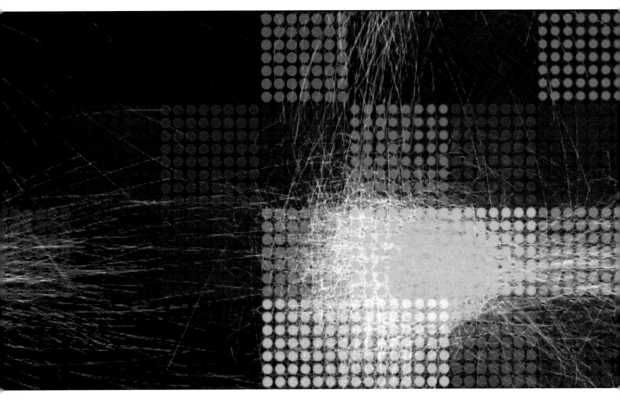

Figure 6.1: Abstract visual score of *Antipode* (one video frame), 2012. Photo courtesy of the Artist.

Figure 6.2: *Antipode* performance, 2012: Ian Whalley at Waikato University in New Zealand (left); Rebecca Cernec and Ivan Zavada at Sydney Conservatorium of Music in Australia (right). Photo courtesy of the Artist.

Figure 6.3: Sydney Conservatorium of Music Computer Music Lab connecting to Syneme studio in Calgary for the first network music collaboration experiment, 2011. Photo courtesy of the Artist.

Figure 6.4: Syneme Studio in Calgary — Han Lu playing *erhu* (left), Ivan Zavada coordinating the network music performance from Sydney, Australia (right), 2011. Photo by Kenneth Fields.

rhythm, dynamics, timbre, texture, emotion, etc. The various combinations of different audiovisual elements in multiple layers generate unlimited creative potential.

The aforementioned experiments led to a new work entitled *Antipode,* a network music composition exploring audiovisual correspondences and the relationship between non-symbolic, abstract visual representations and sonic and musical gestures reflecting the artistic and abstract mind. The concept relies on a grid whereby corresponding column and row coordinates trigger specific image and sound combinations. The projected video shows a combination of one row with one column. The musician performs the associated row/column combination read on the visual score throughout the duration (between 30–60 seconds) of the video projection for that particular section. When the image flickers or becomes very bright and saturated in colour, the musician is to extend and exaggerate the techniques and dynamic range of the improvisation to match the mood of the image. The combination of images ranges from colours (yellow, green, red, orange and blue) with textured images with small, medium and large dots and various pixelated motion graphics.

The main components of the work rely on the control of visual combinations (selected via iPad) in tandem with accompanying layers of sound to suggest various modes corresponding to the imagery. These sonic modes are determined by the textures provided by the sound processing systems – in particular, granulation effects based on the sounds of pre-recorded violin tracks. The violin techniques were performed by Ivan Zavada and the resulting timbres and textures included long sustained notes, note slides or *portamenti, pizzicati, col legno battuto,* sliding double-stops played in tremolo and sliding *pizzicati.* All these techniques covered a wide pitch and dynamic range to allow versatility in the choice of techniques for the granulation process. All pre-recorded techniques were edited into one audio file to facilitate playback and audio scrubbing (playing back an audio file at various speeds and at different locations) to quickly find appropriate textures for the video. The granulation process was a live response to visual stimuli generated by the video, which in turn created sound cues or aural stimuli for the control of visual effects. The sound and visual components were treated independently by two performers and therefore created a creative dialogue and negotiation within the aesthetic realm. In addition, a third network performer, Ian Whalley (on wind synthesiser), interpreted audiovisual combinations through musical improvisations and the result was fed back from New Zealand into the network nodes at the Sydney Conservatorium of Music studios to be mixed and redistributed worldwide for the '121212 UpStage Festival of Cyberformance'. The performance gradually became a synchronous musical dialogue as the performance feedback provided from the network performer in New Zealand also influenced decisions made locally resulting in ever changing visual and sonic textures. This implies a form of communication feedback, locally, regionally and over the network, to accommodate and encourage interactive improvisation, thus creating interdependency throughout the performance.[12] All influences, local and remote multi-sensorial feedback, were essential in creating a synergy with a heightened concentration by all performers to interpret several audiovisual streams.

The work *Antipode* involved software development, composition and a performance, which relied on acoustic instruments, processed sounds, software instruments, mixers, microphones, cameras, digital controllers and a tele-conferencing unit. The system components and network distribution for *Antipode* consisted of three computers running patches programmed in Max, a graphical multimedia programming environment. The network configuration involved the management of point-to-point sound connections, or sound nodes, internet and ethernet routers and a connection manager called JackTrip. The performance system coordinated control messages through protocols such as MIDI (Musical Instruments Digital Interface) and OSC (Open Sound Control) and was operated through controllers such as the Lemur MIDI/OSC application for iPad with a custom template to transform visuals through virtual sliders and a grid of toggle buttons to create multi-layered visual textures. A MIDI device was used to control sound effects, processing and mixing between network performer and locally generated sounds. The sound streams were mixed live and processed further via a digital audio workstation (DAW) such as Logic to add effects such as compression, equalisation, delay and reverb before sending the resultant sound to the concert venues. It is worth noting that tactile latency on the iPad controller in OSC causes an additional performance latency compared to acoustic instruments, where the performers' gestures translate directly into sound. The combination of different types of control systems creates a multi-parametric dependency requiring new skill sets, optimised gestural control solutions and mapping strategies to generate a unified yet open approach for live audiovisual and network creativity/performance.[13]

Intermedia explorations call for wider expertise and experience with various forms of media. Besides the added layer of complexity of network communication and connectivity between the two remote locations, understanding the technology behind every component of the work was crucial for the rendition of the work *Antipode*. It is not only the media that needs to be understood but the interaction modes between various components which in themselves become the new focus to achieve interesting musical and aesthetic results, unlike more conventional or traditional performance practice such as solo, chamber and ensemble music for example.[14] The performers involved in the rendition of a traditional piece of music must be attuned to every single change in character, tempo, dynamics, and emotions evoked during a performance. With new performance contexts and technologies, new issues and challenges arise. Network music performance implies that performers are not located in the same acoustic space such as a concert hall. The communication paradigm shifts from a multi-sensorial experience to a reduced performance practice scenario purely relying on abstract interpretations of both sound and visuals. The physical dimension of performance takes a different form and shifts towards a less tangible, virtual, more cognitive and perceptual processing of musical events. The creative aspect of the performance becomes engrained in the rendition of the work, rather than prescribed by a score or set of specific performance instructions.

In recent network music projects undertaken at the Sydney Conservatorium of Music, the initial set-up was a lengthy configuration process to ensure proper connectivity both internally between the different sound nodes and control channels, as well as externally to secure a proper musical and visual exchange between participants in different geographical locations. The remote location of musicians and artistic collaborators disconnects the physical aspect of performance and introduces a new performance topology, which affects the way performance and creative systems are designed. In the case of *Antipode*, because the audience was located in various installation spaces and cafés around the world, there was no one else present in the studios besides the performers in Sydney, Australia, and Hamilton, New Zealand. In this extended performance situation the live aspect of the performance was weakened by the fact there was no direct audience involvement or participation to guide the emotional and aesthetic directions for the performance, blurring the musical response typically found in a concert hall situation, where there is immediate feedback to physical and perceptual elements of a stage presence.[15] The event was a live concert performance, albeit without any direct audience feedback, which made the experience surreal because it was known to the performers that there was a broad listening audience worldwide that could see and hear the live studio performance, but without direct involvement or known recognition or feedback at the end of the performance.

From a creative standpoint, a distinguishing feature of *Antipode* is the coordination aspect, integral to the work through visual cues and dynamic structure states immersing both performers and the audience in a multilayered audiovisual and perceptual realm. What constitutes musical collaboration in this extended sonic space, through network technology, is what redefines the creative aspect of network music performance. The notion of collective creation involves an intrinsic and cooperative dialogue where participants negotiate, modify and share creative content.[16] Coordinating this creative process may involve a conductor, an autonomous agent such as a software algorithm, or a sequence of instructions like a musical score. Moreover, the coordination and instructional aspect may be the essential attribute of the system design, circumventing the need for conducting. Within the latter context, as the performance evolves it becomes less improvisatory and more structured due to the nature of the abstract visual score. Conceptual information in the form of aural and visual cues embedded in the dynamic score and live mixing of sound nodes engenders a synchronised performance response and performance interdependency throughout the creative process.

Network music collaboration and communication is a natural extension of performance practice involving technology and offers new music production avenues and aesthetics. In recent projects undertaken at the Sydney Conservatorium of Music, bringing together multiple audiovisual applications and processing techniques in a cohesive and constructive way was the most challenging aspect of performance preparation. This remained an integral part of the creative process for works such as *Jasmine* and *Antipode*, involving several active participants in different locations. Reflecting this, engaging modes of interaction and networked collaborations give rise to new models of creative activity, shifting the conventional composition paradigm to collective and interdependent processes rather than idiosyncratic and more traditional approaches to musical creativity. This implies a redefinition of roles, processes and aesthetic decisions the composer engages into within a collaborative space. A distinctive feature of integrated system design is the inherent capacity to incorporate unified perceptual mechanisms based on aural and visual cues through dynamic visual scoring and live sound processing. One of the most significant aspects of network creativity resides in the amalgamation of existing electronic music and audiovisual processing techniques to expand aesthetic boundaries of network music performance as a new means to collaborate, share and explore artistic expression by way of integrating network technologies and expanding sonic space.

ENDNOTES

1   Kenneth Fields, 'Syneme: Live', *Organised Sound* 17(1) (2012): 86–95.

2   Ian Whalley, 'Internet2 and Global Electroacoustic Music: Navigating a Decision Space of Production, Relationships and Languages', *Organised Sound* 17(1) (2012): 4–15.

3   David Kim-Boyle, 'Network musics — play, engagement and the democratization of performance' (Proceedings of New Interfaces for Musical Expression Conference, 2008): 3–4.

4   Álvaro Barbossa, 'A Survey of Network Systems for Music and Sonic Art Creation', *Leonardo Music Journal* 13 (2003): 53–59.

5   Zefir Kurtisi, Gu Xiaoyuan and Lars Wolf, 'Enabling Network-centric Music Performance in Wide-area Networks', *Communications of the ACM* (2006): 52–54. David Howard and James Angus, *Acoustics and Psychoacoustics*, (Oxford: Elsevier, 2009).

6   Barbossa, 'A Survey'.

7   Fields, 'Syneme'.

8   Udo Zölzer and Xavier Amatrin, *Digital Audio Effects* (John Wiley and Sons: Sussex, 2002).

9   Gil Weinberg, 'Interconnected Musical Networks: Toward a Theoretical Framework', *Computer Music Journal* 29(2) (2005): 23–39.

10 Jonathon Stock, 'Concepts of World Music and their Integration within Western Music Education', *International Journal of Music Education* 23 (1994): 3–16.

11 Jonathon Stock, 'Three 'Erhu' Pieces by Abing: An Analysis of Improvisational Processes in Chinese Traditional Instrumental Music', *Asian Music* 25 (1993): 145–176.

12 Whalley, 'Internet2'.

13 Whalley, 'Internet2'.

14 Barbossa, 'A Survey'.

15 Roger Mills, 'Dislocated Sound: A Survey of Improvisation in Networked Audio Platforms', *Proceedings of the 2010 Conference on New Interfaces for Musical Expression*, Sydney, Australia (2010): 186–191.

16 Nicolas Makelberge, 'Rethinking Collaboration in Networked Music', *Organised Sound* 17(1) (2012): 28–35.

INVENTIONS AND
RECOMBINANT POETRY

BILL SEAMAN
OTTO E. RÖSSLER

INVENTIONS AND RECOMBINANT POETRY

BILL SEAMAN
OTTO E. RÖSSLER

Network and friendship are two sides of the same coin—if understood in the right fashion

Conversations in friendship—like this one in the footsteps of Pask

Complexity and Meta-complexity—a consequence of this essay?

Complexity is built out of elements—like these? The future is still waiting ...

Surprise is complexity's offspring—Casti's science of surprise is emergence in a nascent state

The AHA phenomenon is the other source and can arise in total isolation as a breakout into laughter

But it was invariably triggered by previous interaction—either horizontally or vertically

'Make everything as simple as possible but no simpler', said Einstein

From Recombinant Poetics to The World As Interface—a miraculous bisociation

Cartesian Meditations and the nightmare of assignment—if misunderstood, as can happen so easily—everything good is also an assignment condition

The pressure of death is contributing to the joy of invention—if this does not sound too depressed

The pressure of the Now—is a new insight

An almost immoral ceiling of freedom—is the creation of personhood

Recombinant Informatics is a synonym to informatics—the genetic code is a miraculous example of Schrödinger's aperiodic crystal—in recursive evolution

Re-thinking The Database is a version of philosophy, not just a branch

Interflow architectures / new architectonics / the architecture inside Jupiter is waiting to be discovered / since there is life on Jupiter

Meta-invention is the invention of a new type of inventions / like recombinant modular code

Thrown together into a new pattern through a rocking motion—are the lyrics of a lullaby

The invention of sleep—the deepest of history? The long sleep of death was just shown to be bridged by the new infinite cosmology—in which every Everett world is implemented

The moment out of nothing in which the computer hallucinates the understanding of the other as if he were not a computer

Thrown together via music—Nada Brahma and Lieselotte Heller's eigen tone —the acoustic mind's eye—Gerold Baier's fresh morning melody each day

The melody of pauses made palpable as a beautiful experience Craig Tattersall's dusted loops—interruptions of a single tone conveying the creator's voice—and Yves Klein's symphony of silence

Poetics by definition is love—angry poignancy can be a special case

The pure sound of intonation is a carrier—a whole layer of invention—sometimes holy

Displacement illuminates placement and defines it—like in Kurt Vonnegut's 'D.P.' story in Welcome to the Monkey House

The invention of Immensity through cortical hormones

Computers as dissipative open systems—unless of the quantum type

Interfaciology—a linguistic vehicle—a form of theology including antitheology

Input + Functionality = Output as an emergent system and invention generator [human in the loop]—hybrid morphing

Invention generator—physical level, informational level and meaning level

A poem and a pun—compressed information like Freudian condensation

If you can logically describe it—having to do with writing computer code

The invention of picture-switching in the brain—many off-screen screens waiting for their debut

The iceberg of associations—and the 'relevant popping jolt' which miraculously appears as the invention of invention

Biomimetics—to find the physiological multi-value-logic code—or physical logic

Nonsense logic—and its deepest level—the Aion is a child on the throne playing where Heraclitus meant the creator by the word Aion which means eternal

The Logic of the body—is love even with eroticism not excluded

Neosentience—the new ethics of the third millennium—after answering the question of how inventions should be weighed in terms of their potential misuse—if one has the choice to release it or not—but others might decide differently—hence the Benevolence Engine has its own positive vibrational effects

The Insight Engine—the outside copy (an enigma)—reverse engineering research processes and devising new computational ones

The one billion-euro project—This system would be the perfect example of human–computer symbiosis—the first human | computer reciprocal relationship

A brain finds a well-formed mathematical problem that needs the level of a computer that brings the same problem to life—humans as an element in a loop

The insight engine as a tool—neosentience is a higher-order learning system, all the way up to point omega—iterating our book—the book becomes a process of recombinant informatics and can only be fully understood through these transdisciplinary cross-overs and collisions

Iterative learning Systems—brave pupils contributing to the next order of brave pupils—the eusocial revolution is a different one again

A book about post-Darwinian evolution—Techno-Lamarckism recursive evolution, life is such a machine—not quite so ...

We are using our brain in a symbiotic machine along with other human persons

Mind maps—midday naps—Einstein said the bed, the bus and the bath—Archimedes saw the water replace the golden crown—and Poincaré remembered having had his best idea while stepping onto a bus—something is making a connection which one controls—one also works on problems while asleep—in the morning things can have switched from elusiveness to clarity—a crystallisation point that congeals in the ocean of the subconscious

Associative bridging is criminal in a playful sense—one must be careful not to link out to the point of no return

Our work is a conversation that has looped over and over again—a focused wandering—similar to the dérive—Debord

Bridging languages between disciplines—destroys creativity [?]— but on the contrary contributes to creativity. Negotiating negotiating

Collapsing texts—resurrected love—immortality there are so many loops in the metaphysical domain

Mapping texts, embracing trees and rhizomes

New goals for human invention, sustainability and tree-huggery—the deep love of nature

Comparing texts is a big science—after numerical methods and semantic approaches—what is it that humans do? Multimodal pattern flows are licking at our feet—computers will help us feel this tender lapping when they are taken seriously as centres of optimisation and benevolence—multimodal benevolence

Colliding texts—the new enfolding—remembering code as a text—an example is yellow—the sensation of colour—understanding the multimodal requires benevolence and the force fields of attraction and repulsion

Chance within ranges—without any chance—this is loading the dice

Meaning—becoming—Having been—the world generator lets meaning become becoming

Biomimetics—robocryptics—erobotics—robohaptics—tickling fingers

Bio-abstraction—death-reality overcome—an inversion—when did the sensation start? And when did meta-sensation start?—the invention of metaphysics—it is already with the young child—it is not innate, it's an invention—it's the precondition of culture

Spawning neologisms to reflect on these recombinatoric inventions come from nowhere—paleontologies of inventions are neologisms

Open Order Cybernetics—Closed Order Monasteries—Spiralling loops of desire against closed labyrinths of machinic despair—the opposite of when the glass ceiling is shattered by entering a new architecture of thought

The paradox unfolds—the light explosion of the new personhood—the metaphysics of desire brought to life in the machine—this metaphysics is inaccessible to the machine until it gains personhood—the source of culture

The invention of the discussion of good before and after the arising of evil from the outside

Folds and Bridges and Loops and Nestings—Origami and architectures and cascades of innovation

Removals and reunifications—revivals and goodbyes—Here is to the end of the world as we know it—There is looking at you kid

Using one thing for an intention that was not considered in its design

Opening actual windows with engines of streetcars and desire —

Making a pun reflecting on a film of water or on Hal's daisy

Dancing puns and wordplay as long as it is not illegal

Humour and sadness transcended in ...

Meta-bisociation and down to earth breakfast

Human knowing and machinic becoming vs. metabolic becoming and synthetic genetics as a third possibility

When a molecule becomes a message—A new linguistics of media elements and processes

Transcending, or not quite reaching, the digital—the mixed analog-digital future

The acceptance of noise—the transcontextual is born of navigating noise and hyperchaos

The invisible noise when the noise is you—cooling of the sensors does not help in this case

The inseparable identity of generative systems with themselves

Emergence and immersions—self-reflection is always confined to the One

We can only imagine the other

Even if it is machinic

Only a machine can connect to the other—Or be uploaded—Or become one with the other

Emergent apps and apes

The moment when the Orangutan's intelligence mutated

Emergent Intention Matrix—EIM—Enabling one to shift the attention of polysensors

EIM nodal structure for polysensing—The construction of friendly worker bees—as a benevolent blanket drone forming the arch dress—to protect against nanotechnological intrusion

Recombinant Modular Code (new approaches to self-generated programming)

Collapsing texts—resurrecting love—the fruit of desire

Mapping texts, embracing trees and rhizomes and recursions

Comparing texts is a virgin science

Colliding texts—the new enfolding—a neomachinic metaphor

Chance within ranges—better than no chance—Clearer than refraction in water and in mind

Biomimetics—robocryptics

Bio-abstraction—replacing the body one organ at a time—death-reality inversion?

Where the living connects with the nonliving—in the brain

Neologisms—paleontologies inverted—an archeology of bodily media—a variant on variantology—aren't we all just cyborgs?

Open Order Cybernetics—Closed Order Monasteries— the new robotic order—Björk's order—All is full of love

Folds and Bridges—origami and tearings apart—synthetic furies mitigated

Removals and reifications—the workings of miracles

Using one thing for an intention that was not considered in its design—Huxleyan evolutionary ritualisation reproduced artificially

Opening actual windows with computers—in the footsteps of Einstein's imaginary jump

Puns and wordplay as long as it is the language of secret compliments

Humour and sadness in Leonardo's smile

Meta-bisociation and down-to-earth breakfast—right now?

The new linguistics of media elements and processes—unveiled

Generative Systems emulating Emulating

Emergence and immersions—emergent apps and orangutans' intelligence

Emergent Intention Matrix—nodal structure for polysensing and group self-reflection—overcoming the double bind

Recombinant Modular Code—new approaches to Object-based construction

As long as it is sober

As long as it is wide

As long as it is benevolent—this loop in which you are ...

ABOUT THE CONTRIBUTORS

ANDY DONG

Professor Andy Dong is an ARC Future Fellow at the University of Sydney and holds the Warren Centre Chair for Engineering Innovation. His research addresses the central activity of engineering: the design of new products and services. He joined the University of Sydney in 2003 after completing his PhD and postdoctoral training in mechanical engineering at the University of California, Berkeley. He started in the Faculty of Architecture, Design and Planning as a Lecturer and eventually became the Head of Discipline for the Design Lab. In 2012 he joined the Faculty of Engineering & Information Technologies. He is on the Editorial Board of the key journals in design research including *Design Studies, Journal of Engineering Design*, and *Research in Engineering Design*.

BRAD BUCKLEY

Professor Brad Buckley is an artist, activist, urbanist and Professor of Contemporary Art and Culture at Sydney College of the Arts, the University of Sydney. He was educated at St Martin's School of Art, London, and the Rhode Island School of Design. His research, which operates at the intersection of installation, theatre and performance, investigates questions of cultural control, democracy, freedom and social responsibility. He is the editor, with John Conomos, of *Republics of Ideas* (2001) and *Rethinking the Contemporary Art School: The Artist, the PhD, and the Academy* (2009). Buckley has also developed and chaired (with Conomos) a number of conference sessions for the College Art Association, most recently 'The Erasure of Contemporary Memory' (New York, 2011). He has been a visiting artist and professor at numerous institutions including the University of Tsukuba (Japan), the Nova Scotia College of Art and Design University (Canada) and at the Royal Danish Academy of Fine Arts. During 2009, Buckley was a Visiting Scholar at Parsons The New School for Design (New York).

JOHN CONOMOS

Associate Professor John Conomos is an artist, critic and theorist at Sydney College of the Arts, the University of Sydney. He exhibits extensively both locally and internationally. Conomos' video, *Lake George (After Mark Rothko)*, was exhibited at the Tate Modern, London, in February 2009. In 2008 Conomos was represented in the 'Video Logic' exhibition at the Museum of Contemporary Art Australia. In the same year Conomos' commissioned radiophonic essay on Luis Buñuel, 'The Bells of Toledo', for ABC Radio National was also broadcast. Conomos is a prolific contributor to local and overseas art, film and media journals and a frequent keynote speaker and participant in conferences, forums and seminars. In 2000 Conomos was awarded a New Media Fellowship from the Australia Council for the Arts. Conomos' most recent exhibition, 'The Spiral of Time', was shown at the Australian Centre for Photography and was accompanied by a monograph (with Brad Buckley). His research interests include contemporary art, cultural theory, film aesthetics and history, the history of ideas and new media studies.

MARC NEWSON

Born in Sydney, Newson spent much of his childhood travelling in Europe and Asia. He started experimenting with furniture design as a student and, after graduating from Sydney College of the Arts, was awarded a grant from the Australian Crafts Council with which he staged his first exhibition—featuring the Lockheed Lounge—a piece that has now, 20 years later, set three consecutive world records at auction.

Newson has lived and worked in Tokyo, Paris, and London where he is now based, and he continues to travel widely. His clients include a broad range of the best known and most prestigious brands in the world—from manufacturing and technology to transportation, fashion and the luxury goods sector. Many of his designs have been a runaway success for his clients and have achieved the status of modern design icons. In addition to his core business, he has also founded and run a number of successful companies, including a fine watch brand and an aerospace design consultancy, and has also held senior management positions at client companies, including currently being the Creative Director of Qantas Airways.

Marc Newson was included in *Time* magazine's 100 Most Influential People in the World and has received numerous awards and distinctions: he was appointed The Royal Designer for Industry in the UK, received an honorary doctorate from the University of Sydney, holds Adjunct Professorships at Sydney College of the Arts, the University of Sydney and Hong Kong Polytechnic University, and most recently was awarded a CBE (Commander of the Order of the British Empire) by Her Majesty the Queen.

His work is present in many major museum collections, including the MoMA in New York, London's Design Museum and V&A, the Centre Georges Pompidou and the Vitra Design Museum. Having set numerous records at auction, Newson's work now accounts for almost 25% of the total contemporary design market.

Photo: Simon Upton

LEA BARNETT

Lea Barnett is the design director at CampbellBarnett, design partners. She graduated from Sydney College of the Arts, the University of Sydney, in 1991 and from the College of Fine Arts, the University of New South Wales, with an MA in 1993. Barnett has a background in traditional graphic design, arts-based conceptual thinking in sculptural space and new media, and museums. This breadth of experience informs her talent for seeing beyond the brief to interpret a client's real communication issues and deliver perceptive and innovative solutions into architectural space, corporate and marketing communications. Prior to joining CampbellBarnett, Barnett held senior design positions at the Museum of Sydney, the Australian Museum, the Organising Committee of the Olympic Games (SOCOG) and the Art Gallery of NSW. Barnett is a member of both the DIA and AGDA.

PETRA GEMEINBOECK

Dr Petra Gemeinboeck is a Senior Lecturer at the College of Fine Arts, the University of New South Wales. She is also an adjunct Senior Lecturer in the Faculty of Architecture, Design and Planning at the University of Sydney. Gemeinboeck practice crosses the fields of architecture, interactive installation, mobile media, robotics, and visual culture. In her current research, Gemeinboeck collaborates with Dr Rob Saunders to develop robotic interventions as an investigative lens into the co-evolution of humans and machines by deploying autonomous robots, embedded into the architectural fabric of a gallery.

ROB SAUNDERS

Dr Rob Saunders is a Senior Lecturer in the Faculty of Architecture, Design and Planning at the University of Sydney. His primary research interest is the development of computational models of creativity. The development of computational models of creative processes provides opportunities for developing a better understanding of human creativity, producing tools that support human creativity and possibly creating autonomous systems capable of creative activity. His approach to developing computational models of creativity is to develop curious agents and to use these agents to simulate creative systems.

MARI VELONAKI

Associate Professor Mari Velonaki is the director of the Creative Robotics Lab at the National Institute for Experimental Arts, the University of New South Wales. Velonaki is also an Adjunct Associate Professor in the Faculty of Engineering and Information Technologies at the University of Sydney. She has worked as an artist and researcher in the field of electronic art since 1995. Her work creates intellectually and emotionally engaging human–machine interfaces incorporating movement, speech, touch, breath, electrostatic charge, artificial vision, light, text and robotics. In 2009 she was awarded an Australian Research Council Queen Elizabeth II Fellowship (2009–2013) for the creation of a new robot in order to develop an understanding of the physicality that is possible and acceptable between a human and a robot.

DAVID RYE

Associate Professor David Rye is a co-founder of the Australian Centre for Field Robotics, which was established in the School of Aerospace, Mechanical and Mechatronic Engineering at the University of Sydney in 1999. He holds a PhD in Mechanical Engineering from the University of Sydney. He has conducted extensive research in fields related to automation and control of machines, including applied nonlinear control, container-handling cranes, excavation, and autonomous vehicles. Since 2003 he has worked in the field of social robotics, designing and implementing autonomous robots that can interact with people in social spaces. Rye is recognised as a pioneer in the introduction and development of university teaching in mechatronics, having instituted the first Australian BE in mechatronics in 1990.

DAN LOVALLO

Professor Dan Lovallo is at the University of Sydney Business School and is a Senior Research Fellow at the Institute for Innovation Management and Organization at the University of California, Berkeley. His research is concerned with psychological aspects of strategic decisions and has appeared in the *American Economic Review, Management Science* and the *Harvard Business Review*. He received his PhD from the Haas School of Business at the University of California, Berkeley. He is a member of the Editorial Board for the *Strategic Management Journal* and the McKinsey Finance. Dan is currently leading an ARC discovery project titled 'Behavioural Strategies for Selecting Innovation Projects'.

JOHN TONKIN

John Tonkin is a Lecturer in the Digital Cultures program in the Faculty of Arts and Social Sciences at the University of Sydney. He is also an artist and interactive designer and programmer. He exhibits artworks both nationally and internationally. In 1999–2000 he received a fellowship from the New Media Arts Board of the Australia Council for the Arts. Tonkin is currently working on a number of projects that use real-time 3D animation, visualisation and data-mapping technologies. These include 'Strange Weather', a visualisation tool for making sense of life, and 'time and motion study'. Tonkin has worked as an educator in a number of capacities, including lecturing in multimedia at a number of Sydney institutions including TAFE, Metro Screen, UTS and the Australian Film Television and Radio School.

KIT MESSHAM-MUIR

Dr Kit Messham-Muir is a Senior Lecturer in the Faculty of Education and Arts at the University of Newcastle, where he is the Program Convenor of the Bachelor of Fine Art. He graduated from Sydney College of the Arts, the University of Sydney, with a Bachelor of Visual Arts in 1994 and in 2000 was awarded a PhD in Art History and Theory from the University of New South Wales. He was a full-time academic in the Museum Studies unit at the University of Sydney in 2002–2005, and continued to teach at the University as sessional staff in 2005–2007. He has won several awards for his teaching, publishes frequently and works internationally on the StudioCrasher video project. Messham-Muir's research examines the ways in which interpretive practices in galleries and museums impact upon visitors' experiences, particularly by evoking memory and emotion. He is currently working on a book that explores the role of imaging technology in contemporary warfare through the video installation work of Shaun Gladwell, one of Australia's Official War Artist.

ALEX GAWRONSKI

Dr Alex Gawronski is an artist, writer and Lecturer at Sydney College of the Arts, the University of Sydney, from which he also holds a PhD. Much of Gawronski's research is concerned with investigating the gallery as a type of self-generating narrative space in which discourses of representation and power collide. His work has been exhibited at the Museum of Contemporary Art Australia and the Art Gallery of New South Wales. Gawronski is also a co-founding director of the Institute of Contemporary Art Newtown (ICAN).

DAGMAR REINHARDT

Dr Dagmar Reinhardt is a Lecturer in the Faculty of Architecture, Design and Planning at the University of Sydney. Reinhardt's research considers the emerging potential of digital architecture as an exciting challenge, specifically in the interdisciplinary exchange between academic environment, cultural productions, industrial expertise and architectural entrepreneurship. Reinhardt completed her PhD in 2010 at the University of Sydney, holding two full scholarships, IPA and EIPRS. Her research contributes to a contemporary architectural design discourse by looking at the potential of a materiality of duration to foster an architectural capability that can generate 'latent formations' in the design process and sustain performative relations once the design has been materialised.

LIAN LOKE

Dr Lian Loke is a Senior Lecturer in the Faculty of Architecture, Design and Planning at the University of Sydney. Her research and creative practice is interdisciplinary and spans the arts, design and human–computer interaction. She places the lived body at the core of inquiry into contemporary issues and emerging technologies. Her creative works incorporate performance and installation to engage and critique new technologies. Her research interests lie in understanding the lived experience of people interacting with emerging technologies and exploring how to design future products and systems from such understandings.

CHRIS L. SMITH

Associate Professor Chris L. Smith lectures in the Faculty of Architecture, Design and Planning at the University of Sydney. His research is concerned with the interdisciplinary nexus of philosophy, biology and architectural theory. He has published on the political philosophy of Gilles Deleuze and Félix Guattari; technologies of the body; and the influence of 'the eclipse of Darwinism' phase on contemporary architectural theory. Presently, Smith is concentrating upon the changing relation the discourses of philosophy, biology and architecture maintain in respect to notions of matter and materiality and the medicalisation of architecture.

IVAN ZAVADA

Dr Ivan Zavada is a composer, multimedia programmer and designer who lectures at the Sydney Conservatorium of Music at the University of Sydney. His research focus is on the interactive relationship between image and sound within the realm of electroacoustic music. He is currently developing a computer application to represent and generate melodic motifs in three dimensions based on their geometric properties. Zavada's work questions the conceptual nature of music by examining the relationship between concrete sounds on fixed recorded medium and visual elements of abstraction rendered in computer graphics. Zavada is also an accomplished violinist who has performed and recorded with various music ensembles in Canada and the United States.

SEAN LOWRY

Dr Sean Lowry is a Lecturer in the Faculty of Education and Arts at the University of Newcastle and holds a PhD from Sydney College of the Arts, the University of Sydney. He is a Sydney-based artist, writer, academic, musician and producer. Lowry has performed and exhibited extensively both nationally and internationally, presented his ideas at numerous international conferences, and published broadly on topics related to contemporary art and culture. His research ranges from subliminally inflected pop music secretly reintroduced to the market from which it was sampled, to expanded painting and micro-video projects. Lowry's most recent initiative is a global peer-reviewed space for art at the outermost limits of location-specificity.

OTTO E. RÖSSLER

Professor Otto E. Rössler is an Austrian born in Berlin into an academic family. Rössler was awarded his MD in 1966. After postdoctoral studies at the Max Planck Institute for Behavioral Physiology in Bavaria, and a visiting appointment at the Center for Theoretical Biology at SUNY-Buffalo in 1969, he became Professor for Theoretical Biochemistry at the University of Tübingen in 1976. In 1994, he was made Professor of Chemistry by decree. Rössler has held visiting positions at the University of Guelph (Mathematics) in Canada, the Center for Nonlinear Studies of the University of California at Los Alamos (Chaos), the University of Virginia (Chemical Engineering), the Technical University of Denmark (Theoretical Physics), the Santa Fe Institute (Complexity Research) in New Mexico and the University of the Arts in Vienna. He has some 500 scientific publications and believes in visual thinking as being superior to calculating.

BILL SEAMAN

Professor Bill Seaman is currently a Professor at Duke University in the Department of Art, Art History & Visual Studies, focusing on Visual and Media Studies. He is a member of the Duke Institute for Brain Sciences. Seaman was a Visiting Fellow at Sydney College of the Arts during 1990. He was formerly founding head of the Digital+Media Program at Rhode Island School of Design. He earned a Master of Science in Visual Studies in 1985 from the Massachusetts Institute of Technology, Cambridge. In 1999, he received a PhD from the Centre for Advanced Inquiry in the Interactive Arts at the University of Wales. He has taught in a number of institutions in Australia and the United States. He has exhibited in the United States, Europe, Canada, Australia, South America and Japan and has won several international awards: a grant from the National Endowment for the Arts (United States, 1987), the Siemens Stipendium at the ZKM (Zentrum für Kunst und Medientechnologie, Karlsruhe, Germany, 1994), the Ars Electronica distinction prize (interactive art section, Linz, Austria, 1992 and 1995) and first prize at the Berlin Video/Film Festival (multimedia art section, 1995).

BIBLIOGRAPHY

Akhmatova, Anna Andreevna. *The Complete Poems of Anna Akhmatova.* Translated by Judith Hemschemeyer. Brookline: Zephyr Press, 1998.

Anderson, Terry. *The Theory and Practice of Online Learning.* Edmonton: AU Press, 2008.

Barad, Karen. *Meeting the Universe Halfway: Quantum Physics and the Entanglement of Matter and Meaning.* Durham: Duke University Press, 2007.

Barbossa, Álvaro. 'A Survey of Network Systems for Music and Sonic Art Creation', *Leonardo Music Journal* 13 (2003).

Bataille, Georges. *The Accursed Share: An Essay on General Economy. Translated by Robert Hurley.* New York: Zone Books, 1991.

Benjamin, Walter. *On the concept of history* [1940]. New York: Classic Books America, 2009.

Bennett, Jill. 'What is Experimental Art?' *Studies in Material Thinking* 8 (2012).

Bernstein, Basil. *Class, Codes and Control,* London: Routledge, 2003.

—. *Pedagogy, Symbolic Control, and Identity: Theory, Research, Critique.* Edited by Donaldo Macedo, Revised edn, Critical Perspectives Series. Oxford: Rowman & Littlefield, 2000.

Blum, Susan. *My Word! Plagiarism and College Culture.* New York: Cornell University Press, 2009.

Bourriaud, Nicolas. *Relational Aesthetics.* Paris: Presses du reél, 2002.

Broeckmann, Andreas. 'Remove the Controls. Machine Aesthetics', 1997. Accessed May 20, 2013. v2.nl/events/machine-aesthetics.

Brooks, Rodney. *Flesh and Machines.* New York: Pantheon Books, 2002.

Bruce, Mazlish. *The Fourth Discontinuity: The Co-Evolution of Humans and Machines.* New Haven: Yale University Press, 1993.

Bruner, Jerome 'On Cognitive Growth I & II.' In *Studies in Cognitive Growth: a Collaboration at the Center of Cognitive Studies,* edited by Jerome Bruner, Rose R. Olver, et. al. New York: Wiley, 1967.

Buckley, Brad and John Conomos, editors, *Rethinking the Contemporary Art School: The Artist, the PhD, and the Academy.* Halifax: NSCAD University Press, 2010.

Burnham, Jack. 'System Esthetics'. In *Great Western Salt Works: Essays on the Meaning of Post-Formalist Art.* Edited by Jeffrey C. Burnham. New York: George Braziller, 1974.

Butler, Judith. *Bodies That Matter: On the Discursive Limits of 'Sex'.* New York: Routledge, 1993.

Cairns, Susan. 'Mutualizing Museum Knowledge: Folksonomies and the Changing Shape of Expertise', *Curator: The Museum Journal,* 56 (2013).

Cardoso, Carlos and Petra Badke-Schaub. 'Fixation or Inspiration: Creative Problem Solving in Design', *Journal of Creative Behavior* 45 (2011): 66–82.

—. 'The Influence of Different Pictorial Representations During Idea Generation', *The Journal of Creative Behavior* 45, (2011): 130–146.

Carvalho, Lucila, Andy Dong, and Karl Maton. 'Legitimating Design: A Sociology of Knowledge Account of the Field', *Design Studies* 30, (2009).

Chen, Rainbow Tsai-Hung, Karl Maton, and Sue Bennett. 'Absenting Discipline: Constructivist Approaches in Online Learning'. In *Disciplinarity: Functional Linguistic and Sociological Perspectives*. Edited by Frances Christie and Karl Maton. London: Continuum, 2011.

Clark, Andy and David Chalmers. 'The Extended Mind', *Analysis* 58 (1998): 7–19.

Clarke, Arthur C. *Voice Across the Sea*. New York: HarperCollins, 1958.

Cohen, Wesley M. and Daniel A. Levinthal. 'Innovation and Learning: The Two Faces of R&D', *The Economic Journal* 99, (1989).

Colless, Edward 'Superabundance', *Broadsheet: Contemporary Art & Culture*, 33, (2004), 16–18.

Collini, Stefan. Introduction to *The Two Cultures*, by Charles Percy Snow. London: Cambridge University Press, 1998.

Comte, Charles. *Traité de la propriété*, 2 vols. Paris: Chamerot, Ducollet, 1834.

Cope, Bill and Mary Kalantzis. 'The Role of the Internet in Changing Knowledge Ecologies', *Canadian Journal of Media Studies*, 7 (2010): 1–23.

Cowling, David. 'Social Media Statistics Australia—April 2013', *Social Media News*. Accessed 6 May 2013, www.socialmedianews.com.au/social-media-statistics-australia-april-2013/; 'Oceana and South Pacific.'

Csikszentmihalyi, Mihaly. 'Implications of a Systems Perspective for the Study of Creativity'. In *Handbook of Creativity*, edited by Robert J. Sternberg. Cambridge: Cambridge University Press, 1999.

Cunningham, Stuart. *Hidden Innovation: Policy, Industry and the Creative Sector*. Brisbane: The University of Queensland Press, 2013.

Daniels, Dieter and Barbara U. Schmidt. Introduction to *Artists as Inventors, Inventors as Artists*, edited by Dieter Daniels and Barbara U. Schmidt. Ostfildern: Hatje Cantz, 2008.

De Long, Harriet. '9 Evenings: Theatre and Engineering' (manuscript); Series 1. Project files; box 2, '9 Evenings', 1966; Experiments in Art and Technology; Records, 1966–1993; Getty Research Institute.

Deleuze, Gilles and Félix Guattari. *A Thousand Plateaus: Capitalism and Schizophrenia*. Minneapolis: University of Minnesota Press, 1987.

—. *Anti-Oedipus*. Translated by Robert Hurley, Mark Seem and Helen R. Lane. Minneapolis: University of Minnesota Press, 1983.

—. *Kafka: Toward a Minor Literature*. Translated by Dana Polan. Minnesota: University of Minnesota Press, 1986.

—. *What is Philosophy?* Translated by Hugh Tomlinson and Graham Burchell. New York: Columbia University Press, 1994.

Deleuze, Gilles. *Francis Bacon: The Logic of Sensation*. Translated by Daniel W. Smith. London: Continuum, 2003.

—. *The Logic of Sense*. Translated by Mark Lester and Charles Stivale. New York: Columbia University Press, 1990.

DeLillo, Don. *Cosmopolis*. New York: Scribner, 2003.

Dong, Andy. 'The Enactment of Design through Language', *Design Studies* 28, no. 1 (2007).

Downton, Peter. *Design Research*. Melbourne: RMIT University Press, 2003.

Duggan, Maeve and Joanna Brenner, *The Demographics of Social Media Users—2012*. Pew Research Center's Internet & American Life Project: Washington DC, 2013. Accessed 6 May 2013, pewinternet.org/Reports/2013/Social-media-users.aspx.

Dupuy, Jean Pierre. *The Mechanization of the Mind: On the Origins of Cognitive Science*. Translated by M. B. De Bevoise. Princeton: Princeton University Press, 2000.

Edwards, Paul. *The Closed World: Computers and the Politics of Discourse in Cold War America*. Cambridge: MIT Press, 1997.

Eisenman, Peter. 'From Object to Relationship II: Giuseppe Terragni—Casa Giuliani Frigerio', *Perspecta* 13–14 (1971): 38–61.

Fields, Kenneth. 'Syneme: Live', *Organised Sound* 17(1) (2012): 86–95.

Frascari, Marco *Monsters of Architecture: Anthropomorphism in Architectural Theory*. Maryland: Rowman & Littlefield Publishers Inc., 1991.

Gawronski, Alex. 'Artist Run = Emerging—a Bad Equation', *Art Monthly Australia*, 257 (2013).

—. 'Show me Good Art: Biennales and Artist Directed Culture', *Broadsheet*, 39 (2013).

Gemeinboeck, Petra and Rob Saunders, 'Urban Fictions: A Critical Reflection on Locative Art and Performative Geographies', *Digital Creativity* 22, (2011).

Gowlett, John A. J. 'Artefacts of Apes, Humans, and Others: Towards Comparative Assessment and Analysis', *Journal of Human Evolution* 57 (2009).

Graham, Beryl and Sarah Cook. *Rethinking Curating: Art after New Media*. Cambridge: MIT Press, 2010.

Greenfield, Adam. *Everyware: The Dawning Age of Ubiquitous Computing*. Berkeley: New Riders, 2006.

Greenhow, Christine, Beth Robelia and Joan Hughes, 'Research on Learning and Teaching with Web 2.0: Bridging Conversations', *Education Researcher*, 38 (2009).

Grosz, Elizabeth A. 'Future of Space'. In *Architecture from the Outside: Essays on Virtual and Real Space*. Edited by Elizabeth Grosz. Cambridge: MIT Press, 2001.

Groys, Boris. 'On the Curatorship'. In *Art Power*. Edited by Boris Groys. Cambridge: MIT Press, 2008.

—. 'What is Art/Everyone has to be an Artist/After the End of ass Culture/Egypt as a Quick Run'. In *The Future of Art: A Manual*. Edited by in Ingo Niermann. Berlin: Sternberg Press, 2011.

Guattari, Felix. *Chaosmosis*. Bloomington and Indianapolis: Indiana University Press, 1995.

Guédon, Jean-Claude. *In Oldenburg's Long Shadow: Librarians, Research Scientists, Publishers and the Control of Scientific Publishing*. Washington DC: Association of Research Libraries, 2001.

Haque, Usman. 'The Architectural Relevance of Gordon Pask', *Architectural Design* 77:4 (2007).

Hayles, N. Katherine. *How We Became Posthuman: Virtual Bodies in Cybernetics, Literature, and Informatics*. Chicago: University of Chicago Press, 1999.

Heartney, Eleanor. 'In Praise of Uncertainty'. In *Critical Condition: American Culture at the Crossroads*. Edited by Eleanor Heartney. Cambridge: Cambridge University Press, 1997.

Heiser, Jörg. 'State of the Art', *Frieze* 133 (2010).

Held, Richard and Alan Hein, 'Movement-produced Stimulation in the Development of Visually Guided Behavior,' *Journal of Comparative and Physiological Psychology* 56(1963): 872–876.

Hennessy, John L. 'The Role of Creativity and the Arts in a 21st-Century Education: President Hennessy's annual address to the academic council'. April 20, 2006. *Stanford News*. Accessed 21 April 2013, news.stanford.edu/news/2006/april26/hentext-042606.html.

Holloway, Jr., Ralph L. 'Culture: A Human Domain', *Current Anthropology* 10, no. 4 (1969).

Horkheimer, Max and Theodor W. Adorno. *Dialectic of enlightenment*. London: Allen Lane, 1972.

Howard, David and James Angus, *Acoustics and Psychoacoustics*, Oxford: Elsevier, 2009.

Husserl, Edmund. *Cartesian meditations: An introduction to phenomenology* Kluwer Academic, 1960.

Hyatt-Johnston, Helen and Virginia Ross, 'Introduction', in *The Changing Critical Role of Artist Run Initiatives Over the Last Decade*. Edited by Mark Jackson. Sydney: Firstdraft, 1995, 19–29.

Jansson, David G. and Steven M. Smith. 'Design Fixation', *Design Studies* 12, no. 1 (1991); A. Terry Purcell and John S. Gero, 'Design and Other Types of Fixation', *Design Studies* 17, no. 4 (1996).

Johnson, Steven *Emergence—The Connected Lives of Ants, Brains, Cities and Software* (London: Penguin, 2001).

Joselit, David. 'Painting Beside Itself', *October* 130 (2009).

Kelly, Caleb. *Cracked Media*. Cambridge: MIT press, 2009.

Kim-Boyle, David. 'Network musics — play, engagement and the democratization of performance', *Proceedings of New Interfaces for Musical Expression Conference*, Genova, Italy (2008): 3–4.

Klüver, Billy and Julie Martin, 'Working with Rauschenberg' in *Robert Rauschenberg: A Retrospective*. Edited by Walter Hopps et al. New York: Guggenheim Museum Publications, 1998.

Kolarevic, Branko and Ali Malkawi. *Performative Architecture: Beyond Instrumentality*. New York: Routledge, 2005.

Krauss, Rosalind. 'Death of a Hermeneutic Phantom: Materialization of the Sign in the Work of Peter Eisenman'. In *Houses of Cards*. Edited by Peter Eisenman. New York: Oxford University Press, 1987, 166–84.

—. *A Voyage on the North Sea: Art in the Age of the Post-Medium Condition*. London: Thames & Hudson, 2000.

Kurtisi, Zefir Gu Xiaoyuan and Lars Wolf. 'Enabling Network-centric Music Performance in Wide-area Networks', *Communications of the ACM* (2006): 52–54.

Lacerte, Sylvie. '9 Evenings and Experiments in Art and Technology: A Gap to Fill In Art History's Recent Chronicles' in *Artists as Inventors—Inventors as Artists,* ed. Dieter Daniels et al. Ostfildern: Hatje Cantz Verlag, 2002.

Lamont, Alexandra and Karl Maton, 'Choosing Music: Exploratory Studies into the Low Uptake of Music GCSE', *British Journal of Music Education* 25, no. 3 (2008).

Lamont, Alexandra and Karl Maton. 'Choosing Music: Exploratory Studies into the Low Uptake of Music Gcse; Karl Maton, 'Knowledge-Knower Structures in Intellectual and Educational Fields'. In *Language, Knowledge and Pedagogy: Functional Linguistic and Sociological Perspectives*. Edited by Frances Christie and James R. Martin. London: Continuum, 2007.

Landau, Marie. *Narratives of Human Evolution*. New Haven: Yale University Press, 1993.

Lecercle, Jean-Jacques. 'The Pedagogy of Philosophy', *Radical Philosophy* 75 (1996).

Lingis, Alphonso. *Foreign Bodies*. New York: Routledge, 1994.

Lynn, Gregg. *Animate Form*. Princeton: Princeton Architectural Press, 1998.

Makelberge, Nicolas. 'Rethinking Collaboration in Networked Music', *Organised Sound* 17(1) (2012): 28–35.

Mamykina, Lena, Linda Candy and Ernest A. Edmonds, 'Collaborative Creativity', *Communications of the ACM* 45 (2002): 96–99.

Manovich, Lev. *The Database, Language of New Media*. Cambridge, Mass. MIT Press, 2001.

Marr, David. *Vision: A Computational Investigation into the Human Representation and Processing of Visual Information*. New York: Henry Holt and Co., Inc., 1982.

Maton, Karl. 'The Wrong Kind of Knower: Education, Expansion and the Epistemic Device', in *Reading Bernstein, Researching Bernstein*. Edited by Johan Muller, Brian Davies, and Ana Morais. London: Routledge, 2004.

Maturana, Humberto R. and Francisco J. Varela, *Autopoiesis and Cognition: The Realization of the Living*. Boston: D. Reidel Pub. Co., 1980.

—. and Francisco J. Varela, *The Tree of Knowledge: The Biological Roots of Human Understanding,* Boston: Shambhala, 1987.

Mauss, Marcel. *The Gift: Forms and Functions of Exchange in Archaic Societies*. London: Cohen & West, London, 1969.

Menges, Achim and Michael Hensel, *Versatility and Vicissitude*. London: Wiley, AD, 2008.

Merleau-Ponty, Maurice. *Phenomenology of Perception*. London: Routledge, 1962.

Mesulam, M. Marsel. 'From Sensation to Cognition', *Brain* 121, no. 6 (1998).

Mills, Roger. 'Dislocated Sound: A Survey of Improvisation in Networked Audio Platforms', *Proceedings of the 2010 Conference on New Interfaces for Musical Expression,* Sydney, Australia (2010): 186–191.

Myin, Erik. 'An Account of Color Without a Subject?' *Behavioral and Brain Sciences* 26 (2003): 42–43.

Neubauer, Bruce J., Richard W. Hug, Keith W. Hamon and Shelley K. Stewart. 'Using Personal Learning Networks to Leverage Communities of Practice in Public Affairs Education', *Journal of Public Affairs Education*, 1(2011): 9–25.

O'Regan, John Kevin and Alva Noë, 'A Sensorimotor Account of Vision and Visual Consciousness', *Behavioral and Brain Sciences* 24 (2001): 939–972.

—. 'What It Is Like to See: A Sensorimotor Theory of Perceptual Experience', *Synthese* 129 (2001): 79–103.

O'Sullivan, Simon. 'From Aesthetics to the Abstract Machine: Deleuze, Guattari, and Contemporary Art Practice' in *Deleuze and Contemporary Art*. Edited by Simon O'Sullivan et al. Edinburgh: Edinburgh University Press. 2010.

Pallasmaa, Juhani. *The Eyes of the Skin: Architecture and the Senses* Chichester: John Wiley and Sons, 2005.

Papastergiadis, Nikos. *Cosmopolitanism and Culture*. Cambridge: Polity Press, 2012.

Penny, Simon. 'Bridging Two Cultures: Toward an Interdisciplinary History of the Artist-Inventor and the Machine-Artwork' in *Artists as Inventors—Inventors as Artists*. Edited by Dieter Daniels et al. Ostfildern,: Hatje Cantz Verlag, 2002.

—. 'Representation, Enaction and the Ethics of Simulation', In *FirstPerson: New Media as Story, Performance, and Game*. Edited by Noah Wardrip-Fruin and Pat Harrigan. Cambridge: MIT Press, 2004.

—. 'Trying to Be Calm: Ubiquity, Cognitivism, and Embodiment.' In *Throughout: Art and Culture Emerging with Ubiquitous Computing*. Edited by Ulrik Ekman. Cambridge: MIT Press, 2012.

Perez-Gomez, Alberto. 'Monsters of Architecture', book review, *Journal of Architectural Education* (46/1, September 1992): 60.

Petersen, Tanya 'Somnambulant Sensing', *RealTime* 100 (2010).

Petra Gemeinboeck et al., 'On Track: A Slippery Mechanic-Robotic Performance', *Leonardo* 43 (2010).

Pickering, Andrew. 'Cybernetics and the Mangle: Ashby, Beer and Pask', *Social Studies of Science* 32 (2002).

Plautz, Dana. 'New Ideas Emerge When Collaboration Occurs', *Leonardo* 38 (2005).

Protevi, John. 'Deleuze and Wexler: Thinking Brain, Body, and Affect in Social Context'. In *Cognitive Architecture*. Edited by Deborah Hauptmann et al. Rotterdam: 010 Publishers, 2010, 168–84.

Provan, Alexander. 'All for One', *Frieze*, 153 (2013).

Rahim, Ali. 'Contemporary Processes in Architecture', *Architectural Design* 70 (2000).

Rainie, Lee, Joanna Brenner, and Kristen Purcell. *Photos and Videos as Social Currency Online*. Washington DC: Pew Research Center's Internet & American Life Project, 2012. Accessed 1 May 2013, pewinternet.org/Reports/2012/Online-Pictures.aspx .

Raunig, Gerald. 'Creative Industries as Mass Deception'. In *Culture and Contestation in the New Century*. Edited by Marc James Léger. Bristol: Intellect Press, 2011, 94–104.

Reinhardt, Dagmar and Lian Loke, 'Not What We Think: Sensate Machines for Rewiring'. In *Proceedings of the 9th ACM Conference on Creativity & Cognition*. Sydney, Australia: ACM, 2013: 135-144.

—. 'GOLD (Monstrous Topographies)—Exploring Bodies in Complex Spatiality: Trespassing, Invading, Forging Body(ies)', *International Journal of Interior Architecture + Spatial Design*, Spring issue II (2013).

Root-Bernstein, Bob et al. 'Artscience: Integrative Collaboration to Create a Sustainable Future', *Leonardo* 44 (2011).

Rosati, Lauren and Mary Anne Staniszewski, eds, *Alternative Histories: New York Art Spaces 1960 to 2010*. Cambridge: Exit Art NY and MIT Press, 2012.

Rushkoff, Douglas. *Present Shock: When Everything Happens Now*. New York: Current, Imprint of Penguin Group, 2013.

Sadovnik, Alan R. 'Basil Bernstein (1924–2000)', *Prospects: The Quarterly Review of Comparative Education* XXXI, no. 4 (2001).

Sarett, Lewis H. 'Research and Invention', *Proceedings of the National Academy of Sciences* 80, no. 14 (1983).

Saunders, Rob and John Gero. 'A curious design agent: A computational model of novelty-seeking behaviour in design'. Paper presented at the *Sixth Conference on Computer Aided Architectural Design Research in Asia*. CAADRIA, 200).

Saunders, Rob. 'Towards a computational model of creative societies using curious design agents'. In *Engineering Societies in the Agents World VII*. Edited by Gregory O'Hare, Alessandro Ricci, Michel O'Grady, and 0. Dikenelli. Berlin: Springer-Verlag, 2007, 340–353.

Seaman, Bill. 'The Hybrid Invention Generator', *Leonardo* 38, no. 4 (2005).

Shanken, Edward. 'I Believed in the Art World as the Only Serious World That Existed', Interview with Billy Klüver, in *Artists as Inventors—Inventors* as Artists. Edited by Dieter Daniels et al., Hatje Cantz Publishers, 2008.

Siemens, George. 'Connectivism: A Learning Theory for the Digital Age', *International Journal of Instructional Technology and Distance Learning,* 2 (2005).

Silvera Tawil, David, David Rye and Mari Velonaki, 'Interpretation of the Modality of Touch on an Artificial Arm Covered with an EIT-based Sensitive Skin', *International Journal of Robotics Research* 31(2012).

Slager, Henk. *The Pleasure of Research*. Helsinki: Finnish Academy of Fine Arts, 2012.

Smith, Chris L. 'The Problem of Writing Architecture: Knots, Restraint and Pleasure', *Journal for Cultural Research,* 10 (2), (2006): 185–97.

Smith, Steven M. and Julie Linsey. 'A Three-Pronged Approach for Overcoming Design Fixation', *The Journal of Creative Behavior* 45, (2011).

Snow, Charles Percy. *The Two Cultures*. London: Cambridge University Press, 1998.

Stein, Morris. 'Creativity and Culture', *The Journal of Psychology* 36 (1953).

Steiner, George. *Grammars of Creation*. New Haven: Yale University Press, 2002.

Stewart, John R. Olivier Gapenne, and Ezequiel A Di Paolo. *Enaction: Towards a New Paradigm for Cognitive Science*. MIT Press: Cambridge, 2010.

Stock, Jonathon. 'Concepts of World Music and their Integration within Western Music Education', *International Journal of Music Education* 23 (1994): 3–16.

Stock, Jonathon. 'Three 'Erhu' Pieces by Abing: An Analysis of Improvisational Processes in Chinese Traditional Instrumental Music', *Asian Music* 25 (1993): 145–176.

Taleb, Nassim. *The Black Swan: The Impact of the Highly Improbable*. London: Penguin Books, 2007.

Thomas, Nigel. 'Are Theories of Imagery Theories of Imagination? An Active Perception Approach to Conscious Mental Content', *Cognitive Science* 23 (1999).

Tyas Cook, Edward. *The Life and Works of John Ruskin*. London: Methuen, 1983.

Varela, Francisco J. Evan T. Thompson, and Eleanor Rosch. *The Embodied Mind: Cognitive Science and Human Experience*. MIT Press: Cambridge, 1993.

Velonaki, Mari, David Silvera Tawil and David Rye. 'Engagement, Trust, Intimacy: Touch Sensing for Human–Robot Interaction', *Second Nature,* 2 (2010): 102–19.

—. 'Fish-Bird: Cross-Disciplinary Collaboration', *MultiMedia, IEEE* 15 (2008).

—. 'Shared Spaces: Media Art, Computing, and Robotics', *Computers in Entertainment* 6, no. 4 (2008).

—. 'Sharing Spaces: Risk, Reward and Pragmatism', in *New Constellations: Art, Science and Society*. Sydney: Museum of Contemporary Art, 2006.

Velonaki, Mari, Stefan Scheding, David Rye and Hugh Durrant-Whyte. 'Shared Spaces: Media Art, Computing and Robotics', *ACM Computers in Entertainment* 6 (2008).

Verwoert, Jan. 'Apropos Appropriation: Why Stealing Images Today Feels Different', *Art & Research: A Journal of Ideas, Contexts and Methods,* 1 (2007).

Viswanathan, Vimal and Julie Linsey. 'Examining Design Fixation in Engineering Idea Generation: The Role of Example Modality', *International Journal of Design Creativity and Innovation* 1 (2013).

Ward, Thomas B, Merryl J. Patterson, Cynthia M. Sifonis, Rebecca A. Dodds, and Katherine N. Saunders. 'The Role of Graded Category Structure in Imaginative Thought', *Memory & Cognition* 30, (2002).

—. 'Structured Imagination: The Role of Category Structure in Exemplar Generation', *Cognitive Psychology* 27, no. 1 (1994).

Ward, Thomas B. and Cynthia M. Sifonis. 'Tosh Demands and Generative Thinking: What Changes and What Remains the Same?', *The Journal of Creative Behavior* 31 (1997).

Weinberg, Gil. 'Interconnected Musical Networks: Toward a Theoretical Framework', *Computer Music Journal* 29(2) (2005): 23–39.

Wentworth Thompson, D'Arcy. 'On the Theory of Transformations, or the Comparison of Related Forms' in *On Growth and Form*. Cambridge: Cambridge University Press, 1917.

West, Ruth, Jeff Burke, Cheryl Kerfield, Eitan Mendolowitz, Thomas Holton, J. P. Lewis, Ethan Drucker, and Weihong Yan. 'Both and Neither: In Silico V1.0, Ecce Homology', *Leonardo* 38, 4 (2005).

Whalley, Ian. 'Internet2 and Global Electroacoustic Music: Navigating a Decision Space of Production, Relationships and Languages', *Organised Sound* 17(1) (2012): 4–15.

Wiener, Norbert. *Cybernetics: Or, Control and Communication in the Animal and the Machine*. New York: The Technology Press, John Wiley & Sons, 1948.

—. *Invention: The Care and Feeding of Ideas*. Cambridge: MIT Press, 1993.

Wilde, Oscar. 'Preface' in *The Picture of Dorian Gray*. New York: Simon & Schuster, 2005[1891].

Yakeley, Megan. *Deconstruction for Everyone: The Myth Exploded*. Msc diss., University of East London, 1992.

Youmans, Robert J. 'The Effects of Physical Prototyping and Group Work on the Reduction of Design Fixation', *Design Studies* 32 (2011).

Zhang, Jianwei. 'Toward a Creative Social Web for Learners and Teachers'. *Educational Researcher* 38 (2009): 276.

Zickuhr, Kathryn and Aaron Smith. 'Digital Differences: While Increased Internet Adoption and the Rise of Mobile Connectivity Have Reduced Many Gaps in Technology Access over the Past Decade, for Some Groups Digital Disparities Still Remain' (Pew Research Center's Internet & American Life Project: Washington DC, 2012), 5. Accessed 30 April 2013, pewinternet.org/Reports/2012/Digital-differences.aspx.

Zölzer, Udo and Xavier Amatrin. *Digital Audio Effects*. Sussex: John Wiley and Sons, 2002.

LIST OF FIGURES

INDEX